T0331681

"A reader will finish *SMALL SCIENCE* with a big-time lesson in biology, and because of Platt's writing, may not even realize what has happened. We even have his artwork displayed in a real gallery and poetry ... to augment his tales of what he calls small science but what in fact turned out to be an enviable career as a multi-faceted and highly productive scholar."

John Janovy, Jr.
Varner Professor Emeritus, University of Nebraska-Lincoln, USA

"*SMALL SCIENCE* is a big book. A frank memoir of Tom Platt's life as a distinguished parasitologist committed to engaging undergraduates in real research, this book is both a detailed celebration of some amazing organisms and a tribute to the scientists and students who come alive in studying them. Platt's voice is intimate, reflective, thoughtful, funny, and focused; and his story will reward anyone interested in science, higher education, and the life of learning in our time."

Patrick E White
President Emeritus, Millikin University, USA

"Against the depictions of modern science on the grand scale — formidable and dehumanizing — this gem of an autobiography reminds us of the small science that still persists: its research problems and disciplinary communities, and lives and work of those who practice it as teacher-researchers in liberal arts colleges."

Christopher S Hamlin
Professor of History, University of Notre Dame (IN), USA

"Thomas Platt's recollections through a fascinating life and career are loaded with interesting historical events, details on diverse biological phenomena emphasizing parasitology, and numerous adventures in meeting interesting people and travels throughout the world. Platt candidly relates the joys, frustrations, and incredible serendipity that can occur in a teaching and research career. The book is a wonderful celebration of our field and an exemplary source of advice for parasitologists, students, and academic researchers."

Ramon A Carreno
Professor of Zoology, Ohio Wesleyan University, USA

SMALL
SCIENCE

Baracktrema obamai and Other Stories of a
Life in Parasitology & Higher Education

SMALL
SCIENCE

Baracktrema obamai and Other Stories of a
Life in Parasitology & Higher Education

Thomas R Platt

Saint Mary's College, Notre Dame, USA

World Scientific

NEW JERSEY · LONDON · SINGAPORE · BEIJING · SHANGHAI · HONG KONG · TAIPEI · CHENNAI · TOKYO

Published by

World Scientific Publishing Co. Pte. Ltd.

5 Toh Tuck Link, Singapore 596224

USA office: 27 Warren Street, Suite 401-402, Hackensack, NJ 07601

UK office: 57 Shelton Street, Covent Garden, London WC2H 9HE

Library of Congress Cataloging-in-Publication Data

Names: Platt, Thomas Reid, author.

Title: Small science : Baracktrema obamai and other stories of a life in parasitology &
    higher education / Thomas Reid Platt, Saint Mary's College, Notre Dame, USA.

Description: New Jersey : World Scientific, [2021] | Includes bibliographical references and index.

Identifiers: LCCN 2021015320 | ISBN 9789811239144 (hardcover) |
    ISBN 9789811239809 (paperback) | ISBN 9789811239151 (ebook for institutions) |
    ISBN 9789811239168 (ebook for individuals)

Subjects: LCSH: Parasitologists--United States--Biography. | Parasitology--Research.

Classification: LCC QL31.P525 A3 2021 | DDC 591.7/857092 [B]--dc23

LC record available at https://lccn.loc.gov/2021015320

British Library Cataloguing-in-Publication Data

A catalogue record for this book is available from the British Library.

For any available supplementary material, please visit
https://www.worldscientific.com/worldscibooks/10.1142/12339#t=suppl

Desk Editor: Shaun Tan Yi Jie

Typeset by Stallion Press
Email: enquiries@stallionpress.com

Printed in Singapore

For Kathy — who else?

Contents

Preface

Why write a memoir? We are curious, and that curiosity extends to lives we will never experience unless somebody is willing to share theirs. Many accounts in this genre provide catharsis, an opportunity to unburden the writer of personal demons experienced and overcome in their lives. The reader will find nothing of that ilk here. My life is devoid of any ill-treatment by family, teachers, clergy, or anyone else. Others relate personal accounts of historical interest and provide insights not available in traditional scholarly treatments. Again, not here. The events I describe are personal or involve people rarely in the public eye. What I share is my life in higher education and what I call small science. Similar accounts could be written by hundreds of my colleagues who labored in liberal arts colleges now and in the past. Their stories would be at least as engaging, if not more so, than mine.

I am a parasitologist: not a psychotherapist, parapsychologist, physical therapist, or other occupations with nominally similar-sounding names. I study parasites, and I have for nearly half a century. When asked what I do by someone I've just met, and they stare blankly when I share my occupation, I follow with, "If you take your dog to the vet(erinarian) to get it wormed, I study the worms." Responses range from the understated, "Well, I guess somebody's got to do it," to facial expressions implying a note of disgust. I have been on the receiving end of both. However, either reaction almost invariably turns to curiosity. Most people have encountered parasites: ticks on a dog, fleas on a cat, pinworms in a (friend's) child, or something more exotic, and they have questions. Nobody wants to be in a car wreck, but most folks slow down to look. Same for parasites.

I didn't start with the notion of writing a memoir. The idea developed gradually. I started with the chapter, "In Defense of Parasites." Why? I named a trematode in honor of President Obama — *Baracktrema obamai*. During my 15 minutes of fame, I was dismayed by the negative comments that appeared in the press, not about the president or me, but about parasites. I wanted to say something on their behalf. The *Washington Post* published an op-ed I wrote that hinted at an extended argument no newspaper would consider. I started there. Then I wrote a second piece detailing the events that occurred during the brouhaha that erupted following the introduction of *Baracktrema*. That led to several other essays about discoveries and events, both scientific and personal, I thought might be of interest. Then I began to contemplate the improbable path that led to my life in Parasitology. Bookstores are replete with volumes regaling the lives of paleontological prodigies, but none about parasitologists. I thought I had a story to tell — except the prodigy part. This is it.

The primary motivation for this project? To see if I could do it. That was my initial justification for pursuing a PhD. I was not a stellar student in high school and early college years. I discovered that passion later. As I made my way through a Master's program, I developed a love of science and learning. I wondered if I had the intelligence and persistence to complete a doctorate. Later in life, I wondered if I had the wherewithal to write a book — any book. The answer was always, "No." Once I retired and completed several projects from my working life, I was at loose ends. I didn't have anything to do. I don't play golf or tennis or have any other hobbies to occupy the vast chunks of time I faced in my golden years. I did this instead.

I wrote the bulk of what follows from memory. I reviewed some of my publications to nail down specifics of research projects completed years ago; and, Google Maps helped confirm, or correct, recollections of distances traveled. I also relied on journals written during stays in Australia and Malaysia for specific information — the only times in my life I journaled. I am sure I got some things wrong. I strongly suspect that some of the people I mention may remember things differently. If I offended anyone, I apologize in advance — except for D.A. Azimov and Colleen. You will meet them both in due course.

I recently celebrated my 72nd birthday and have been retired for six years. If my 18-year-old self could see where life would take him in the next half-century, he would be stunned. The geezer has been happily married for 48 years to a wonderful and supportive woman. He has two sons he loves and respects, and those feelings are reciprocated. Both graduated from college debt-free and have good jobs. He has two amazing grandsons whom he sees regularly. And he and his wife have enough money for a comfortable retirement. My 18-year-old self would learn that, despite a few bumps in the road, he can look forward to a life far better than he could have ever imagined.

None of what I accomplished could have happened without the love and support of my wife, Kathy. Few spouses would have put up with the temporary storage of "poop" samples in the fridge or finding a 15-pound snapping turtle staring up at her from the utility sink when she went to wash her hair. These are a few eccentricities she endured during our marriage.

Almost every graduate student includes a tribute to their significant other in the Acknowledgments section of their thesis. I wrote the following over 40 years ago as I was poised to complete my PhD at the University of Alberta:

"Finally, I wish to thank my wife, Kathy, for her understanding and love. Her support during the past four years has been an invaluable commodity. Repayment will take a lifetime."

My debt is still outstanding.

1. The Parasite and the President[68]

Fame is finally only the sum total of all the misunderstanding that can gather around a new name.

— Rainer Maria Rilke

On September 8th, 2016, a Thursday, the *Journal of Parasitology* issued a press release announcing the publication of an article describing a new genus and species of parasite named in honor of the President of the United States, Barack Obama. I discovered the worm, a digenetic trematode, while on sabbatical at the Universiti Putra Malaysia in 2008. It inhabits the blood vessels of the lungs of two species of turtle endemic to Southeast Asia: *Cuora amboinensis* (the Asian box turtle) and *Siebenrockiella crassicollis* (the Black Marsh turtle). As the discoverer of this new organism, I had the honor of the christening, bestowing the name for the new creature according to the rules established by Carl von Linné, or Linnaeus. The name given to this new addition to Earth's bestiary? *Baracktrema obamai*.

Why the lag of eight years between the discovery of the new worm and its formal description? I fully appreciated the importance of my find almost as soon as I saw it. Nothing I had seen in the past quarter-century was remotely similar. It was a new genus and species.

When my stay in Malaysia concluded, I returned to Saint Mary's and began staining specimens and mounting them on slides for microscopic examination. My initial analysis of the "lungworm" confirmed what I had suspected from the beginning. It belonged in the family Spirorchidae

---

*Superscripts refer to publications listed on pp. 257–263.

(these are the turtle blood flukes, or TBFs, and will loom large as this story unfolds); however, it differed from all the other genera described. I began attempting to elucidate the form of the various organs and organ systems and measuring them for formal publication. I was excited!

The basic morphology of trematodes is well known, and I was eminently familiar with the variations present in TBFs. The worm's anterior end was straightforward: an oral sucker surrounded the mouth; a short esophagus led to a single cecum or intestine that ran almost the entire length of the body before terminating near the posterior end of the worm. Typically, trematodes have two blind cecae, but a single cecum evolved independently on numerous occasions across the families of flukes. The vitellarium, an organ that produces nutrients for the developing egg, surrounded the cecum. The testis was elongate, and the ovary compact. All of these structures were clearly visible in my specimens. Near the posterior end of the animal, a cirrus sac was present. The cirrus sac houses an eversible cirrus (analogous to the penis but turns inside out during copulation — ouch!). Most trematodes are hermaphrodites possessing both male and female reproductive organs. The genital pore, the site for both copulation and egg release, was located near the worm's posterior tip. The big problem was between the ovary and the genital pore. The area housed a complex array of ducts, tubes, and sacs I could not decipher.

Over the next six years, I examined these worms between my teaching duties, supervising undergraduate research, and working on other research projects. I was never satisfied I fully understood how all the tubes and sacs connected and their role in this fascinating creature's reproductive life. I was stumped. For one of the few times in my professional career, I could not work out a trematode's anatomy, and I wouldn't publish something I was only guessing at. I planned to retire at the end of the 2014–15 academic year and didn't imagine these specimens would yield their secrets to me.

On the other hand, not seeing this through to completion was not an option. It just wasn't going to be me who carried it across the finish line. I had to find somebody smarter, with better equipment, and the knowledge to complete the investigation.

The choice was easy. Stephen A. "Ash" Bullard, Auburn University, is a generation behind me in age and a generation ahead of me in knowledge and technique. Ash is an expert on the Aporocotylidae (fish blood flukes), which bear many similarities to their cousins found in turtles. My first encounter with Ash was an email request for TBFs for comparison with the fish parasites that formed the basis of his PhD studies with Robin Overstreet at the Gulf Coast Research Laboratory in Ocean Springs, Mississippi. I was happy to oblige. Ash and I talked at parasitology meetings on several occasions over the ensuing years. I admired his intelligence, dedication, and the utter joy he exuded about his work and life. He was the perfect choice to crack the problem I was unable to solve. Ash's response to my inquiry was an enthusiastic and unqualified "Yes!" Then he surprised me by handing the work over to a graduate student interested in turtle parasites.

I was skeptical, but Ash assured me the student showed great potential. His name was Jackson Roberts. I trusted Ash, and I sent him my specimens and all the literature I collected on TBFs over the years. I hoped it would speed the process and save them the trouble of accumulating the hundreds of references already in my possession. With a sigh of relief, I packed all the material in several boxes and shipped them south. I turned to the task of cleaning out my office and laboratory in preparation for retirement.

Occasionally, in films and on television, a person is shown retiring from their job after many decades of service. They walk past offices and cubicles saying goodbye to colleagues carrying a single cardboard box with a few pictures, plaques, and maybe a plant peeking over the top edge. I don't know if this portrayal is accurate, but leaving academic life is orders of magnitude more challenging. During my 45 years as a graduate student and faculty member, I accumulated over 1,000 books, 5,000 reprints (individual copies of articles), files on students, rough drafts of manuscripts, and voluminous correspondence. There were also thousands of microscope slides of worms from research projects, not to mention vials of worms (mostly nematodes) not typically mounted on slides. Deciding their fate would take the bulk of my last sabbatical during the fall of 2014. I had to cross-check information on the slides with my records, and label them for deposit in an accredited museum. I examined files individually to determine what to keep, what to recycle, and what to shred, as many contained potentially sensitive information. In the recent past, student grade reports still included Social Security numbers!

The books were the hardest to let go. I don't know any academic of my generation who doesn't love books. Books mark the history of our lives, our development as scientists and scholars. Many people take them home, but our house was small and already overburdened with books. I planned to continue to do research on a limited basis after retirement, so anything related to those projects stayed. What to do with the rest posed a problem. I wanted to put them in the hands of people who would use them.

The sorting process was slow because of the memories many of the volumes invoked. I could recall when, where, and why I purchased most of them. I was erasing my past one volume at a time. I kept about 100 and decided which friends, colleagues, and former students might like the rest. I found homes for the vast majority of them, and I trust they will serve their new owners well.

As the semester wore on and I was nearing the end of the big "sort," I checked with Ash to see how Jackson was doing. Within days, Ash sent a photograph of Jackson holding multiple sheets of 8 × 11-inch paper taped together, forming a 3 × 4-foot canvass with the preliminary drawing of our worm. Unfortunately, Jackson was behind the sheet, so I couldn't put a face to the name. The pace of activity picked up over the next month, and I received the first draft of the manuscript describing our, as yet unnamed, new genus and species. I was relieved to see that my interpretation of the byzantine network of tubes and sacs was about 90% correct. Jackson clearly and convincingly sorted out the rest. It finally made sense.

After some fine-tuning, the manuscript was ready to be sent out for review; however, the worm still did not have a proper Linnaean binomial. Although I would not be the first author on the paper — that honor was Jackson's — I did, as the person who discovered this creature, have "naming rights." My choice was to name it for a relative. The name? *Baracktrema obamai*.

Why? Why name a new parasite after President Obama? My cousin, Doug Toot, and his wife Lola are serious amateur genealogists and discovered our family connection to the 44th president through a gentleman who resided in Pennsylvania in the late 1700s — George Frederick Toot. He is my 4th great-grandfather and President Obama's 6th, making us 5th cousins twice removed. The primary rule for naming new species is that the binomial must be unique. No other animal can have the same name, the name must be in Latin (or Latinized), and the two names must agree in number and gender. I described 30 new species during my career, and naming them was always a chore. I know nothing about the structure of Latin words, their gender, declension, and such. However, if you name a new species after a person, the only requirement is the addition of an -i at the end of their name for a male or -ae for a female honoree. I used this convention to name over a dozen new species after family and friends as related in later chapters. Since President Obama was part of my "extended" family, I followed a tradition established early in my career.

Having a species named after you is an honor. I have two species and a genus named for me. I was touched my colleagues thought my contributions to the field sufficient to warrant public recognition. I voted for Obama twice and felt he did an admirable job as president during his two terms. Naming this unique and beautiful organism would be, in my mind, a tribute to his legacy.

I finally screwed up my courage and sent Ash my proposal. The initial response from Auburn was lukewarm. Jackson grew up in Tennessee, and many of his relatives were not fans of Obama. Both Ash and Jackson finally indicated their assent, but I sensed a level of discomfort. Ash is a friend. Although I didn't know Jackson personally, I didn't want to force either of them to do something they might find problematic, either personally or professionally. I told them if they didn't want to use the name for any reason, I would change it. I didn't have an alternative in mind, but I would think of something. After a brief interlude, they both decided, "What the hell, let's see what happens." We sent the manuscript to the *Journal of Parasitology* for review and possible publication.

The reviews were positive and recommended publication. One reviewer suggested we should not name the organism after a prominent politician without permission. I expected *Baracktrema obamai* might raise some eyebrows. Hell, I was *hoping* it would. Ours wasn't the first organism named in Obama's honor. It wasn't even the first parasite. None of them caused a kerfuffle as far as I knew; why would one more? I did not attempt to contact the White House. Better to seek forgiveness than ask permission.

In April, Kathy and I traveled to Ft. Mill, South Carolina, for the meeting of the Southeastern Society of Parasitologists. My primary reason for attending was to meet Jackson and the other members of Ash's lab. Jackson was a delight. A large, bear-like young man several inches taller and more than a few pounds heavier than I. We immediately bonded and fell into conversation about our backgrounds in and outside the discipline. Jackson attended college in Tennessee and played baseball in high school and college. We had a great time over the 2 ½-day meeting, which made the 1600-mile round trip more than worthwhile.

The April and June issues of the *Journal of Parasitology* passed with no sign of our paper. It had to be August. In the middle of the month, I received an e-mail from Peter Burns, the liaison between Allen Press and the *Journal of Parasitology*. The journal would issue a press release heralding the publication of *Baracktrema obamai* (Figure 1). A few days later, I received a second e-mail from Peter with a series of questions regarding my motivation for naming a parasite after the president. I was puzzled because the formal description of a new organism contains a short section entitled "Etymology," which explains the derivation of the chosen name. We clearly indicated my familial connection to the president and were naming it in his honor. There could be no doubt we had no intention of disparaging Mr. Obama. I didn't realize my answers (and those of my co-authors) would be crafted into a press release.

Most of the questions centered on the negative view many people had of parasites, and was this really an honor? I thought this a strange question coming from a journal devoted to the study of these organisms. I suspect they wanted to make sure I was on record stating my motives were pure. They probed the family relationship and any significance to the parasite's Malaysian origin (the answer was No).

There was one question I thought was a bit strange: Does something about the new species, especially its physical characteristics, remind you of Obama? I thought about this for a moment and wrote: "I should note this seems the equivalent of when Barbara Walters asked her interviewees what kind of animal or tree they would be — a little silly." With that preamble, I concluded, "The worm is long, thin, and cool as hell!"

Kathy and I planned a trip to New York City for 8–11 September earlier in the year. It was a working vacation. We both love Broadway, and I planned on doing some research in the archives of the American Museum of Natural History for a paper on a dispute between several prominent

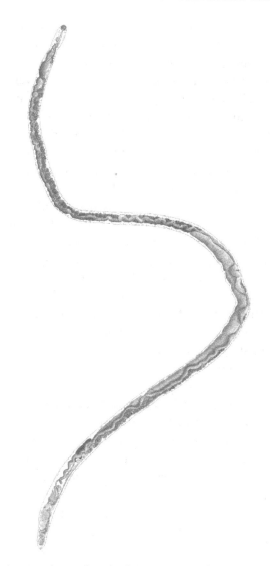

Figure 1. *Baracktrema obamai* from: Roberts, J.R., T.R. Platt, R. Orélis-Ribeiro, and S.A. Bullard. 2016. New genus of blood fluke (Digenea: Schistosomatoidea) from Malaysian freshwater turtles (Geoemydidae) and its phylogenetic position within Schistosomatoidea. *Journal of Parasitology* 102: 451–462. (Illustration by J.R. Roberts. Reprinted with permission.)

parasitologists at the beginning of the 20th century. Kathy would spend time with a high school friend, Karen Pontius, who lives in the city, and we looked forward to attending plays in the evening.

I learned the press release would appear while we were in NYC. Not an ideal time. I don't have a cell phone. I have nothing against technology, and I am not a Luddite. I don't need one. I spend most of my days either in my office at Saint Mary's or at home. There are landlines in both places. When I travel, Kathy is with me, and she has a phone for directions and emergencies. Only this time, she wouldn't be. I would be at the museum, and Kathy would be touring The Big Apple with Karen. I gave the journal my wife's number as my contact, and Kathy had the library's number. An inelegant solution to the problem, but it was the best I could do.

September 9th was a beautiful day, warm and a bit breezy befitting late summer. The stroll from our rental on the upper eastside through Central Park to 79th and Central Park West was a glorious way to begin an eventful 48 hours. I arrived at the museum when they opened, stowed my backpack in a locker as required, and introduced myself to the librarian who would assist me during my visit. I gave her the information I needed for my project, and she headed for the stacks.

I arranged my supplies: a pen, mechanical pencil, paper, and a tablet computer (to photograph documents if necessary). I wanted to examine the letters and manuscripts of Horace Wesley Stunkard, a former research associate of the museum and faculty member at New York University whom I met at my first parasitology meeting over four decades earlier. I was investigating a controversy regarding the early history of the family Spirorchidae: the same family now containing the new genus and species, *Baracktrema obamai*.

I began plowing through the boxes of material the librarian delivered from the stacks. Because of the limited amount of time available, I made judgments of what to examine in detail rather quickly. Dr. Stunkard led a fascinating life. Born in Iowa in the late 1880s, he attended Coe College and earned a PhD at the University of Illinois under Henry Baldwin Ward, the father of American Parasitology.

Horace volunteered for service in World War I as part of the American Expeditionary Force. I had to forego examining documents related to his military service and other honors he received. I hoped to unearth correspondence with Dr. Ward and two other scientific luminaries; George W. MacCallum, a Canadian physician turned parasitologist, and his son, William G. MacCallum, a noted physician and scientist who did parasitological research early in his career. After perusing hundreds of documents in dozens of files and folders, I came up empty. Nothing in the museum archives shed additional light on the controversy surrounding these men. I was tired and disheartened.

In mid-afternoon, my world turned upside down. The librarian told me my wife called and needed to speak to me immediately. Fortunately, I was the only person working in the archives, and the staff granted access to their phone. Kathy wasn't in full panic mode but close. She received calls from reporters who wanted to talk about the article. She had names and numbers for the *Associated Press* and *Philadelphia Inquirer*. I jotted them down. We chatted briefly, and I replaced the handset in the cradle.

My first call was to the *AP*. Most of the information requested was in the press release, but I guess journalists need direct quotes to justify their salary. As soon as I finished with the *AP*, I called the *Inquirer* and repeated the process. Both focused on the fact that I named a parasite for our current (and generally popular) president; parasites are regarded as among the lowest forms of life (obviously, I did not concur). Was this really considered an honor? I reaffirmed my admiration for President Obama and highlighted our familial connection. I shared that I named a parasite (from the eye of a turtle) for my father-in-law and conveyed my life-long dedication to the field and my love of the organisms. I felt I acquitted myself reasonably well.

Kathy, Karen, and I met for dinner. We ordered, and I provided a recap of the interviews. We had purchased tickets to see *The Marvelous Wonderettes* playing at the Kirk Theater on 42nd Street, and the conversation turned to other topics as we ate. During dinner, I silently weighed my options. Should I attend the play or let Kathy and Karen go while I returned to our rental to see if there were any more inquiries I needed to address? My laptop was there, and I knew I had internet access.

We came to New York to see plays. When dinner concluded, we headed to the subway and arrived at the theater about 45 minutes before curtain. In something close to a miracle, my tablet (a new and little-used acquisition) decided to connect to the theater's Wi-Fi. My inbox contained a dozen messages regarding *Baracktrema*. Among them: *CBS* radio, the *Los Angeles Times*, and *The Chronicle of Higher Education*. I also had a rather unpleasant message from a "troll" who felt I was being disrespectful to the president and made some rather graphic suggestions as to what I could do to myself, all of which were physically impossible.

Since there was ample time before the play started, I borrowed my wife's cell phone and called *CBS*. I did a brief interview and turned to the e-mail from the *LA Times*. The reporter sent a list of questions for me to answer. I am not terribly adept at "thumb typing," but I managed a response and pushed "send." I contacted *The Chronicle* and made arrangements for a phone interview the following morning.

There were congratulatory notes from friends, colleagues, and former students who saw reports of our discovery in news outlets here and abroad. I began to wonder if I had done something incredibly stupid. I never sought the limelight. I spent most of my days alone in my office and laboratory. I hoped for some modest recognition, but a few friends suggested *Baracktrema* was going "viral." I wasn't prepared for what was coming.

Saturday was beautiful. We planned a trip to Brooklyn to visit the son and daughter-in-law of dear friends from South Bend, Bob and Ann Cope. Jonathon and his wife, Liza, had adopted a boy and girl born to sisters from Jamaica. We wanted to meet the children and headed across the East River. I had a second objective in mind. The carousel from my hometown amusement park, the now-defunct Idora Park, had been purchased, moved to Brooklyn, and restored to its original condition. On an earlier visit to the city, Kathy, Karen, and I went to ride it only to find the attraction was open every day except Tuesday, and it was Tuesday! All I could do was look at those magnificent horses and carriages through the glass enclosure on the banks of the East River and wait for another day. Today was that day.

Before leaving for Brooklyn, I spoke with the reporter from *The Chronicle*. Her approach to the subject was a bit different. She was interested in reviewing all the flora and fauna named in honor of President Obama. I also received a rather frantic note from Gwen O'Brien, the Media Relations Director at Saint Mary's. She caught wind of what was happening and wanted to coordinate my interactions with the press.

We made arrangements to meet when I returned, and Kathy, Karen, and I headed across the East River.

Jonathon and Liza's apartment consisted of four rooms: bathroom, kitchen/dining room, living room, and bedroom, all in a line, and close quarters for four people. With all the paraphernalia required for two infants and the addition of three visiting adults, there was barely room to move. The babies were beautiful, and we spent time catching up.

After 45 minutes or so, Kathy shared our news with the new parents. Liza stopped dead and said, "Oh my god, that was you? I read about it on *The Skimm!*" I had never heard of *The Skimm*. Liza explained it was an online service providing a synopsis of the news for folks too busy to read more traditional forms of reporting. Each day they published a quote of the day headlining their posts. The selection for September 9th? *It's long. It's thin. And it's cool as hell. — A scientist on why he named a parasite after President Obama. Welcome to the world, Baracktrema obamai.* We were surprised the news of my discovery penetrated the lives of these slightly frazzled, new parents. We said our goodbyes, rode the carousel, had lunch and returned to Manhattan. That evening we attended an excellent performance of *Waitress* on Broadway. The next afternoon, Kathy and I flew back to South Bend.

On Monday, I met with Gwen O'Brien to discuss the rapidly developing story of *Baracktrema*. The story *had* gone viral and appeared in media outlets around the globe. Most made light of the "squirmy honor" but clearly indicated I was sincere in my tribute to President Obama. Conservative papers and bloggers were not as kind, noting the gesture was fitting as in their minds, Obama was a parasite — or worse. In this age of social media, "trolling" is a part of daily life for many people. However, my experience was almost nil as I had no online presence: no Facebook, Twitter, or Instagram. I received two e-mail messages from liberals condemning my action.

During our meeting, Gwen shared that KABC radio in Los Angeles wanted to do a live interview the following morning and asked if I was interested. I said, "Yes." I was curious that Gwen hadn't mentioned the *South Bend Tribune*, our local paper, or any local television stations. She indicated more interest in national and international coverage. I argued we should do everything possible to get Saint Mary's name in front of the local community as we suffer from living in the shadow of the Golden Dome, i.e., the University of Notre Dame. Gwen agreed to contact local reporters. She was good to her word, and the local CBS affiliate, WSBT, would send a reporter later in the day.

Kaitlin Connin fit the mold of a modern newscaster: young, attractive, and very bright. Kaitlin arrived with a cameraman and suggested we chat a bit before starting. I gave her a quick tour of the newly renovated Science Hall and shared some personal history and my career in biology. She and her assistant set up in one of the new laboratories, positioned me on a stool, and checked the lighting and sound levels. The interview was professional and straightforward. As Kaitlin was packing to leave, she indicated the story would air at 6 pm. I missed the live broadcast due to a prior commitment.

When Kathy and I returned home, I fired up the computer and searched for the clip online. I couldn't find it, but I did locate the transcript. I was pleased with the flow until I got to the section on our shared ancestor, the connection that prompted me to name the new worm after President Obama. Kaitlin's article quoted me saying our common ancestor's name was George Frederick Smith, not George Frederick Toot. I nearly fell out of my chair! How could she have made such a horrific mistake? Or had I, in the stress of the moment, misidentified my 4th great-grandfather? I searched again for the video and couldn't believe my ears when my video doppelganger uttered the name "George Frederick Smith." I sent Katlin an e-mail noting my faux pax. Her reply? "Well, it isn't the worst thing that ever happened in broadcast journalism. I'll correct it in the print edition." A charming young woman.

The following morning, I waited in my office for the radio interview with KABC. At some point, I realized I didn't know the ideological leanings of KABC or who might conduct the interview. Some of the press accounts by right-wing media were less than kind, and I thought, "Oh crap, this could be really, really bad." Again, thanks to the internet, I located KABC online and was relieved to hear the mix of news, sports, and humorous banter associated with mainstream drive-time radio. A producer called and gave me some tips about what to expect. I heard the introduction and was live in Los Angeles. The conversation went smoothly; the on-air personalities had some fun at my expense and reined me in when I drifted into professor-speak. Six and a half minutes later, it was over. I said nothing untoward and, unexpectedly, enjoyed the attention.

The press reports were snowballing; however, I was distressed by the derogatory nature of many of them. Even the articles reporting our naming as an honor to the president used modifiers like "dubious" or "squirmy" to indicate the public's negative view of parasites. I decided to write a short piece with the working title of "In Defense of Parasites." The words came easily, and within an hour, I had a 600-word essay. The question was, what next? This was probably my only shot at having *my* voice in the national press, so I went with "Go big, or go home." It had to be either the *New York Times* or the *Washington Post*. The *Times* had done little with the story, while the *Post* published a substantial article with a picture of Obama. I went with the *Post*. I had zero expectation I would get a reply, let alone an op-ed in the paper of Bradlee, Woodward, and Bernstein. I searched the paper's website for the appropriate editor and sent an e-mail explaining who I was, and asked if they would be interested in a short piece from my perspective. I was stunned when I heard back from Mike Larabee expressing interest in reading my "piece" but no guarantee to print it.

Publication of a daily newspaper moves at warp speed compared to its academic counterparts. I was used to months, or more, from submission to print because the information in a scientific paper has relevance for years or decades. The life span of many news articles is the blink of an eye by comparison. The next day I received word the *Post* accepted my essay. After a thorough edit, the final draft entitled *"I named a parasite after Barack Obama. It was meant as a compliment"* (not *In Defense of Parasites* as I had hoped) appeared on Friday, September 15th, less than three days from submission to print! Friends told me not to read the comments. I read all 100+: the good, the bad, and the vicious. People who disliked Obama continued their ignorant screeds. People who thought I was disrespectful to the president hammered me, but a few kind souls rose to my defense, demonstrating an understanding of what I did and an appreciation for the "beauty of life in all its forms." Their thoughtfulness lifted my spirits.

The week between the news release to the *Washington Post* article was a wild ride. Much to my relief, things settled down. I learned a valuable lesson; be careful what you wish for. I was surprised; however, I hadn't heard from the *South Bend Tribune*. Then one of those coincidences of thought and action occurred. The phone rang. Margaret Fosmoe, the education writer for *Tribune*, was on the line requesting an interview. We arranged to meet at my office early the following week.

On Tuesday, Margaret arrived with a photographer in tow. We spoke for about 45 minutes, and I was photographed holding a drawing of *Baracktrema*. The following Saturday, the article appeared on the front page, below the fold. It was similar to others published over the previous weeks; however, Margaret posed an interesting question none of the other reporters thought to ask. After the standard "Have you heard from the White House?" (I had not), she asked, "What do you think Obama's response was when he heard the news?" I thought for a second and replied, "If he did," the professor said with a smile, "my guess is he shook his head in amusement and moved on to more important things."

I did not receive a call from the president or the White House. I suppose I fantasized it might happen, and I would have been delighted if Obama had reached out. I sent a copy of the *Journal of Parasitology* containing our article to the White House for inclusion in Obama's Presidential Library. I received the standard postcard thanking me for my gift. A card mailed to thousands of people every year who send stuff to the president and first lady: plain, perfunctory, and impersonal.

My 15 minutes were over. Gwen O'Brien shared the results from a service the Saint Mary's employed to follow reports of the college in various media: 200+ mentions, more than any other single event in our history. The paper brought nearly 10,000 unique visitors to the *Journal of Parasitology's* website, more than the next 19 articles combined. In a little over two weeks, I received close to 100 congratulatory e-mails from friends and colleagues, here and abroad. I even got a nod from our campus Security Officers when I stopped by their office to renew my parking tag. And the response from the higher-ups at Saint Mary's? Nothing! Not even an "Atta boy" from the Dean, Provost, or President. Their indifference was baffling, disappointing, and a bit hurtful.

Do I regret my decision to "raise some eyebrows"? No. Nobody was hurt. I had a little fun and brought some attention to parasitology and the organisms to which I devoted the better part of my life. If Obama was facing a re-election campaign, I wouldn't have done it. I *do* have a great deal of respect for the man as president, husband, and father. I would not have done anything to hurt his chances for a second term. The furor died down quickly, and *Baracktrema obamai* is still, I suspect, cycling through snails and turtles in Southeast Asia. However, the illegal turtle trade and habitat degradation may threaten the extinction of both hosts and parasite. Neither of the Obamas will live forever, but their name will as long as we are here, and there are folks like me who are fascinated by these genuinely remarkable and underappreciated organisms.

2. Small Science

Somewhere, something incredible is waiting to be known.
— Sharon Begley

Liberal arts colleges are incubators for training future scientists, producing twice as many students who earn PhDs in science than larger institutions. The faculty at these schools often work with little money or recognition. Most small colleges strongly encourage or mandate student participation in research as a requirement for graduation. In the absence of research grants, faculty must rely on ingenuity to allow students to complete projects that result in papers modeling those in scientific journals. These projects, limited by time and scope, are what I call "small science." My introduction to research in small science occurred at my alma mater — Hiram College (Ohio).

I loved Hiram. I thrived (once I got my act together) in the liberal arts setting. I loved the smallness of the place. I loved that the faculty knew their students and welcomed them into their offices and laboratories. I loved the campus, the traditions, and the opportunity to be a true student-athlete. I also knew, when envisioning my future in the mid-to-late 1970s, I wanted to spend my life in similar circumstances.

Some people know they don't want children. In the same vein, I knew I didn't want graduate students. I knew I would do research; however, I didn't want the responsibility of prepping Masters and PhD candidates for qualifying exams, a thesis defense, and sending them out to an uncertain future in the dwindling job market in higher education. I also didn't want to have to find the money to support them. I didn't have the temperament to thrive in a research university. I fell in love with small colleges, and I fell hard. I would spend my life doing small science.

I once asked my then-Dean at Saint Mary's and dear friend, Dorothy Feigl, for suggestions on an invited presentation I was preparing entitled "The Liberal Arts Experience: Prospects for Parasitologists" at the 1991 meeting of the American Society of Parasitologists. Her response? "It is not a place for failed university faculty." She felt teaching in a liberal arts college couldn't be your second choice, a consolation prize. If research was paramount and teaching an afterthought, you would not be happy, and your students would suffer.

All students at Saint Mary's College in South Bend, Indiana, where I spent the bulk of my academic life, were required to complete a Senior Comprehensive Project as a graduation requirement. While the "Senior Comp" was required for graduation, each department developed a program they felt was best for their majors. They ranged from exams to writing projects to empirical research.

The Department of Biology instituted a research requirement long before undergraduate research became *de rigueur* and my arrival. There was, however, no money allocated explicitly for those projects. Each department member received a few thousand dollars annually to purchase material for classroom and laboratory supplies. After I bought everything necessary to teach lectures and labs, anything left could fund student research.

Each student formulated a hypothesis, designed an experiment, collected and analyzed data, and compared her study to the relevant literature. They wrote papers, presented their results in a symposium setting, and in later years, did a poster presentation as well. I supervised 76 student projects in my 28 years at Saint Mary's. That was the norm. Some of my colleagues had a few more and some less, but we all ended up with roughly similar numbers during those three decades. My students and I generated the ideas, and I cadged money from my teaching budget to pay for them. I engaged in this work because it was fun, or when asked by friends of the benefit to society my research might generate, I replied, tongue planted firmly in cheek, "It keeps me busy and off the streets." I was never awake in the wee hours of the morning during the first week in October waiting for a call from Stockholm with news of a Nobel Prize.

Most of my students investigated aspects of the biology of an intestinal parasite maintained in the laboratory using two types of animals: mice and snails. We developed projects that could be completed in a few weeks at most. Most of them cost less than $200 — small science. The name of the parasite? *Echinostoma caproni.*

Echinostoma caproni is a digenetic trematode (a flatworm) discovered in the intestine of the Madagascar Kestrel (*Falco newtoni*) in 1964 and later found in rodents in Egypt. In the early 1970s, Paul Weinstein, University of Notre Dame, brought it to the United States, and parasitologists have been raising it in laboratories worldwide ever since. Adult worms infect a range of rodents (mice and hamsters work well) and occupy the posterior half of the small intestine. Larvae that emerge from the eggs develop in the snail, *Biomphalaria glabrata.* Mice are easy to maintain in the lab, as are the snails. *Echinostoma caproni* is an ideal organism for student research.

I followed a different path for my personal research — taxonomy. I discovered a love of taxonomic work (the identification and classification of living organisms) during my Master's program at Bowling Green (Chapter 6). My PhD research on the taxonomy and life cycle of *Parelaphostrongylus odocoilei* (Chapter 7) reinforced the idea that those types of investigations could be done on a shoestring and would fit the time constraints of the heavy teaching loads typical in small colleges. All I needed was a microscope, slides and coverglasses, and a few reagents. Taxonomy was, until the advent of molecular biology, almost by definition, small science. While taxonomic work is not held in high esteem by some experimentalists, almost any work in biology requires identification of the animal under study. The names, such as *Parelaphostrongylus odocoilei,* are created by taxonomists. Some scientists, working at the cutting edge of the discipline, look on taxonomy with disdain; however, *all biological research rests on a taxonomic foundation.*

Saint Mary's bought me a new compound microscope shortly after I arrived. I never requested any capital equipment over the next 28 years. I used a dissecting microscope from a teaching lab to conduct necropsies of the turtles I examined. I didn't ask for much because I didn't need much. In 35 years in higher education, I received grants totaling $42,327, and $39,969 of that sum funded a year-long sabbatical trip to Australia. Most of the money provided support for travel and living expenses for a family of four. I am not complaining. It was my choice. I didn't have grant money because I didn't apply for grants.

My pursuit of small science may have cost me my position at the University of Richmond (Chapter 8). If it did, so be it. Fortunately, Saint Mary's didn't seem to care. Faculty were expected to engage in "scholarly activity," broadly defined as a condition for tenure and promotion. There were no requirements for obtaining external grants while I was employed. During my term on the Committee for Rank and Tenure, grants were not a factor in our decisions — a policy I hope never changes.

Might I have been more successful with grants? Let's suppose I applied for and successfully received grants — a big if. Probably, but at a cost I wasn't willing to bear. If you accept money, you must produce results on schedule so you can apply for additional funds. It would have meant more nights and weekends in the lab. I never wanted to be driven by anyone's schedule but my own. I doubt any granting agency would have tolerated a 30-year wait to publish the description of *Testudinema gilchristi* (Chapter 21)! While I rarely worked less than 50–60 hours/week, I controlled my time. If there was something more important in my life, research and writing could wait. I rarely missed any of my kids' games or performances, and I was always home for dinner, to help with homework, or to play catch. Any additional time in the lab would have affected my relationships with my wife and children, a sacrifice I was not willing to make.

Could a newly minted PhD follow the same path to a successful career of teaching and research? My answer is a qualified, "Yes." Most liberal arts colleges encourage, if not mandate, research experiences for their students. If you are determined, persistent, and clever, I believe you can succeed. You will teach because you love it. You will put up with long hours and lower pay than your university counterparts; learn to do all manner of things because there isn't anyone else who will; serve on committees that will leave you pounding your head on the table; counsel students who will frustrate and amaze you; and do research because you can't imagine not doing it.

Finally, why Parasitology? Parasitology was an accident. I didn't choose it; it chose me. Parasites are the core of my story; what are they, and how they kept me enthralled for nearly half a century. Let me share a bit about them and then start my story.

3. In Defense of Parasites

The kids are finally asleep, and you settle in to watch a nature show about predators of the African savannah. You would love to take a photo safari to see this magnificent country and its wildlife for yourself; however, time and money constraints being what they are, you will have to settle for an hour of television. The videographers locate a group of female lions stalking a herd of zebras. The lions single out an aging male inadvertently separated from his companions, and the chase is on. The zebra struggles gallantly but is no match for the claws and fangs of the four determined and hungry predators. He is dragged down from behind and subdued as one lioness grabs him by the throat and patiently suffocates the now helpless equid. It is over in a matter of minutes. While you might feel a moment of remorse for the zebra, that changes to a sense of wonder as a group of cubs emerge from the brush to begin feeding on the prize. While a bit gruesome, this is the cycle of life, and these mothers are providing for their offspring just as you provide for yours.

But this is not your typical documentary. As the narrator's voice continues in the background, the camera zooms in on the feast and seamlessly switches to a microscopic view of small, white sacs peppering the flesh. The commentator softly intones that these are cysticerci or larval tapeworms. They will excyst in the intestine of the lion and grow to a length of approximately one meter. It is a British production after all. Tapeworms acquire energy from the partially digested material in the intestine of their feline host by absorption as they lack a gut. At their peak, each tapeworm will release up to 100,000 eggs per day. The eggs will contaminate the grass on which the zebra herd grazes. Once in the zebra's gut, they hatch, burrow through the intestinal wall and produce the cysticerci that initiated the infection.

The camera cuts to an endoscopic view of the intestine with the ethereal lighting typical of a single beam in an enclosed space revealing a long, undulating ribbon of white against the tan and shadows of the poorly lit interior of the small intestine. And you think: Yuck! Gross! Disgusting!

Let's analyze your reaction. You find the killing and devouring of the zebra by the lion pride less disturbing than the parasites because you are a predator. Unlike the lions in the documentary, you probably haven't killed anything for the dinner table personally, but you do hunt. You hunt pork, free-range chicken, peppers, and arugula at the supermarket. You are in search of the energy necessary to sustain yourself and your family. Vegetarians and vegans are equally culpable. Living organisms die as well to support their energy requirements. It is a distinction without a difference. We all seek energy from something else to grow and maintain our bodies, defend against predators, reproduce and get our genetic information into the next generation.

The tapeworm is doing precisely the same thing, only less violently. It has the same goal as the lion: obtain enough energy to grow to adulthood, avoid being killed in the process, and reproduce. Whether lion, tapeworm, or human, those are the only things that matter. Why is our predation less objectionable than the parasite skimming a bit off the top? To grumble that the parasite is taking something it didn't earn is a bit like a bank robber complaining a pickpocket nabbed his wallet. You judge the parasite from a moral perspective, absolving the lion (and yourself) of blame. You might argue the tapeworm causes disease and harms the lion. Well, the lion didn't do a lot to improve the health of the zebra, nor did you do any favors for the cow that provided your last hamburger.

The bottom line is we all need energy, and it has to come from somewhere (and something) else. The flow is unidirectional. In most ecosystems, energy travels from the sun to plants to herbivores (predators on plants) to carnivores (predators on animals) to decomposers (fungi and bacteria who recycle the dead stuff). You might think the plants and decomposers are benign players, but competition for light and dead stuff can be similarly fierce, although those are someone else's story to tell. We are all taking energy from other living things and, in the process, causing harm to satisfy our own biological imperatives of growth, self-preservation, and reproduction. Nature doesn't give a whit about the morality of the encounter. If you are going to survive, something is going to die.

I admit parasites have a significant "yuck" factor for most people. I have frequently fielded questions from family, friends, and strangers, such as: "Why do they exist?"; "What good are they?" and "Can't we get rid of them?" The moral component of these questions is inescapable, yet none of them is difficult to answer if you stick to evolution and leave morality to philosophers and theologians.

Why do Parasites Exist?

To paraphrase career criminal Willie Sutton's possibly apocryphal answer to the question of why he robbed banks — "That's where the energy (money) is." When you look across the dining room table, you see your spouse and children as unique individuals for whom you care deeply and would do anything to assist in their journey through life. A parasite views the same scene, anthropomorphically, of course, and sees big bags of regularly renewed carbohydrates, proteins, and lipids. In other words, the parasite sees a nearly inexhaustible island of energy. From an evolutionary perspective, it would be much more perplexing if the vast supply of energy bound up in the trillions of free-living organisms on Earth was not exploited in some manner other than predation. If parasites didn't exist, we would have to invent them. As the old saying goes, "Nature abhors a vacuum."

Every cell in every organ in the body of a plant or animal is a potential source of energy for some parasite. Even parasites have parasites. As Jonathan Swift noted over 400 years ago, "Big fleas have smaller fleas upon their backs to bite them, and smaller fleas have lesser fleas and so *ad infinitum*." If there is a source of energy available in Nature, some organism will find a way to exploit it to its own evolutionary ends: either all at once or bits at a time, the grand casino heist vs. skimming in the counting room.

The almost infinite variety of habitats also helps address why there are so many different species of parasites. Current estimates suggest parasites (protozoa, worms, and other multicellular forms) make up well over half the total number of species on the planet. Parasitism has evolved on dozens of occasions. These leaps from free-living to parasitic modes of existence occurred independently in most animal and many plant phyla, not including viruses (which are all parasites) and bacteria. Parasitism is, therefore, the most successful strategy for making a living devised by nature. Each host species is a unique ecosystem comprised of a multitude of niches. Each niche requires a particular suite of attributes, both morphological and chemical, to extract the available energy. Natural selection offers the mechanism to fashion an almost infinite number of solutions (unique species of parasites) to exploit the resources available. This description gives the false impression that there could be an endless number of parasite species, but the number is distinctly finite. Parasites range from extreme generalists to those phenomenally specialized in their requirements.

Toxoplasma gondii, a protozoan parasite that only reproduces sexually in the intestine of felids, can infect almost any warm-blooded animal in its asexual phase. There are estimates that as many as 50% of humans carry evidence of this ubiquitous parasite's past or current infection. Then there is a trematode (flatworm or fluke) only found in the oviducts of a single species of turtle in the upper midwestern United States. It does not infect the oviducts of any other turtle species found in the same locality, let alone a male. I have dissected several thousand animals in my career, and I was always more surprised when I didn't find parasites than when I did. I should note that I was only looking for macroparasites (worms of some sort), so it is possible protozoans were present that escaped my gaze.

The planet is awash with parasites. You see a robin hopping across your lawn in early spring in search of food; I see a bag of worms, in addition to protozoans infecting its red blood cells and lice chewing on its feathers. To most people, my view of the world is less romantic and possibly downright "icky," but I find it fascinating.

What are They Good For?

If we remove the moral component from this question, the answer is simple. Parasites are "good for" exactly the same thing you are good for — making more of themselves. No more, no less. To think otherwise is to misread the tree of life. You and the tapeworm have navigated the competition of natural selection and are successfully exploiting energy sources for your own ends. You and the tapeworm are currently two of millions of terminal twigs on the tree of life doing the same. You are not unique. You are not special. Individuals will die, and species will, with certainty, go extinct. Again, not romantic, but true.

Better Hosts Make Better Parasites, and the Reverse is True

The outcome of a parasite infecting a host falls into one of three categories: 1) the host kills the parasite (not good for the parasite), 2) the parasite kills the host (sometimes bad, but sometimes good for the parasite — not so good for the host), or 3) the two negotiate some middle ground where both manage to carry on and complete their biological imperative (good for everyone). Kill or be killed — or coexist.

The habitat parasites encounter differs from those of free-living organisms in one startling dimension. It evolved to kill them. The vertebrate immune system can recognize and respond to an almost infinite number of foreign molecules; to distinguish self from non-self and destroy it. Sure, free-living organisms face predators, but they are rare, opportunistic, and not typically keyed to detect one and only one target. The immune system has two arms; an innate system directed primarily at a suite of molecular forms common to bacteria and an adaptive system with two complementary components: cellular and humoral responses.

Cellular immunity keys on *your* cells gone wrong — cancer and cells harboring parasites. Your immune system identifies molecular markers, antigens on the surface of the cell it recognizes as rogue and produces killer cells to attack and destroy the renegade if all is working well (hint: it doesn't, as cancer is the 2nd leading cause of death in developed nations and intracellular parasites are common).

Humoral immunity identifies antigens independent of your cells and produces free-floating molecules, antibodies, which bind to these antigens and marks them for destruction. Through a remarkable capacity for genetic rearrangement and combining several interlocking component molecules, the adaptive immune system produces antibodies and killer cells that recognize millions of distinct antigens, even molecules synthesized in the laboratory and never found in nature.

The parasite has several options: damp down the immune response or hide from it in one way or another. The first is akin to reducing the size of the police force — fewer cops, less enforcement of the law. The second is Harry Potter's invisibility cloak. Parasites have evolved both mechanisms to bamboozle the vertebrate immune response and survive in this hostile environment. Damping down the immune system results from stimulating the production of regulatory T-cells (T_{reg}) that naturally calm the immune system when an infection is on the wane. Hiding involves capturing molecules from host cells and covering the surface of the parasite with them to blend into the background and appear to be another host cell or tissue. A third ploy is the "moving target" strategy employed by the African trypanosomes, the causative agent of "sleeping sickness" in large swaths of the African continent. These single-celled creatures live in the bloodstream and produce a coat of molecules visible to the immune system. The host responds and begins to decimate the invaders, but a few of the hoard generate a new and different covering. They feed and reproduce, and the immune system ramps up again. This pattern of molecular change, reproduction, and destruction repeats until the host is exhausted and dies. But no matter. By then, the parasite has been carried to a new host by the blood-feeding tsetse fly. This is a simplistic and incomplete overview of how parasites work their magic to avoid detection and destruction by the immune system. Still, you have to agree these are formidable weapons in the evolutionary arms race.

Host and parasite are engaged in a continuous battle of spy vs. spy. The host is trying to kill the parasite, and the parasite is attempting to avoid destruction. Parasites have shorter generation times and a higher reproductive potential than their hosts, resulting in a greater probability of producing mutations and circumvention of the immune system. If the host eliminates 99% of the parasites present, the remaining 1% will pass the mutation (natural selection in action!) to their offspring and successfully infect the next generation of hosts. If some proportion of the host population dies from the parasitic infection, the genes for a weak immune system will disappear from the gene pool. The remaining hosts will pass their ability to contend with the parasite to their offspring. In many instances, the host and parasite reach some rapprochement and coexist at levels where both meet their biological imperative of growth and reproduction. The parasite may cause something we recognize as a disease condition; however, it is not sufficiently debilitating to prevent the host from getting on with its life. The host marshals some energy to control the parasite but not enough to eliminate it. Better hosts make better parasites, and better parasites make better hosts. Go figure!

The medical community has noticed these abilities and is attempting to employ them in service to human health. Researchers use parasites to damp down the immune system to treat autoimmune diseases (malfunctions where the immune system attacks its own cells) such as irritable bowel disease and multiple sclerosis. If we could figure out how schistosomes (trematodes inhabiting the circulatory system of the vertebrate host) coat themselves with host molecules and render them invisible to immune attack, it might provide a mechanism to reduce the rejection of transplanted organs and make the use of immunosuppressive drugs by transplant patients unnecessary.

Host-parasite interactions have effects at the ecosystem level as well. Hosts adversely affected by a parasite might be slightly less attentive or a hair slower than those who eliminated the invader or reached a better level of accommodation, making them easier targets for predators. The predator gets its meal at a discount. The predator doesn't have to expend as much effort to satisfy its energy requirements. It may get parasites in the bargain, but if the burden of infection doesn't affect its reproductive ability, it doesn't matter. And if the parasite doesn't infect that predator, the hunter gets a little extra protein in the bargain!

The death of the host is not always bad for the parasite. Most intestinal worms (flukes, cestodes, nematodes, and acanthocephalans) infect the final host via predation. The next host ingests the parasite's infective stage when it feeds and is nearly impossible to avoid. The cysticerci in the zebra will only complete their biological imperative if the zebra is killed and eaten by the lion. So it is for many parasites. The death of the host is good for the parasite. And parasites don't always sit idly by "hoping" for the host to meet an untimely end; they can alter their behavior to increase their probability of being lunch. The metacercariae (larval trematodes) of *Dicrocoelium dendriticum* turn ants into zombies that attach to blades of grass, making it more likely sheep will eat them while grazing. The same goes for the acanthocephalan, *Polymorphus paradoxus*. Its larval stage, a cystacanth, alters the behavior of a freshwater arthropod, *Gammarus lacustris*, so they grab the feathers of ducks and are eaten by the fowl while preening. The ubiquitous *Toxoplasma gondii* renders mice, a favorite food of cats, less wary and attracted to, rather than repelled by, the smell of cat urine. The cat gets dinner, and the mouse is dead — long live the parasite.

Why Can't We Get Rid of Them?

At this point, you probably realize parasites are too ubiquitous and clever to eliminate. Most are not problematic to their host and not worth our time and treasure. Many serve a useful role in ecosystems, even if they cause some problems for the host. So the real question is, "Can't we get rid of the ones plaguing humanity?" We should but, so far, no such luck. We did rid the world of smallpox, but as one scientist quipped, "Smallpox was a really stupid virus. It never changed antigenically, and it only infected humans." Despite those shortcomings, the eradication of smallpox was an expensive and heroic undertaking. Polio is next on the list. Despite having effective vaccines, civil wars and distrust of the medical community have hampered efforts to vaccinate people in some parts of the world.

The Carter Center in Atlanta made eradication of the Guinea worm, *Dracunculus medinensis*, a priority. Guinea worm is transmitted when humans drink water containing infected arthropods called copepods. The adult females, which reach a meter in length, are found under the skin in the lower extremities of infected humans. When mature, the worm forms a painful blister in the lower leg or foot. Immersion in water provides temporary relief. When the blister hits the water, it ruptures, and the female worm releases thousands of larval nematodes into the pool or pond. Once free, these tiny worms are eaten by copepods and develop to a stage infective to humans. Contamination of the water supply typically occurs when women (mostly) and children enter the water after walking miles from their homes to fill containers with water for household use, a chore vital to the lives of too many families living in poverty. People are infected when they drink water containing the infected copepods. The worms escape in the digestive tract, migrate out into the body, and mate. The gravid female is ready to release her young about a year later.

Eradication should be easy. When *we* go to the kitchen or drinking fountain, we expect the water we are about to drink to be free of debilitating worms and pathogens. *We* take clean water for granted. Not so for much of the planet. The Carter Center and other charitable institutions have invested heavily in drilling wells and distributing simple filters that provide water free from the copepods carrying the infective larval nematodes. Their success has been nothing less than spectacular. The number of human infections across a broad swath of the African continent has dropped from several million to a small handful in isolated areas plagued by civil war. But now the disease is showing up in dogs, and transmission may continue even if human infections disappear. This makes the probability of eradication more complicated. Clean, uncontaminated water for every person on the planet, something we expect without much thought, would eliminate virtually all human infections of this parasite. The addition of the dog as a reservoir host means eradication is going to take more thought.

In 1985, WGBH (Boston) produced a film entitled *Conquest of the Parasites*. The program focused on the "Great Neglected Diseases," aka parasites, which affected hundreds of millions of people in Third World countries in the tropics. This was the beginning of the molecular revolution in biology. Kenneth S. Warren of the Rockefeller Foundation attempted to enlist the talents of young scientists trained in molecular techniques to take on the challenge of developing cures for some of the most egregious of these parasites. I recall him saying something to the effect of "I am betting these scientists are smarter than a bunch of parasites." They no doubt are. In 2018, I gave a lecture at my *alma mater*, Hiram College, entitled *Are you smarter than a trematode?* The title was a takeoff on Jeff Foxworthy's television show *Are you smarter than a 5th grader?* and designed to draw at least a small audience as I was competing with three other presentations. The answer was, "Of course you are; trematodes are dumb as dirt!" Trematodes and other parasites are not smart, but natural selection has made them clever.

Smart vs. clever was the distinction Dr. Warren failed to identify. The program focused on attempts to develop a vaccine for malaria, which at the time was killing approximately 1.5 million people annually, mostly children under five. The hope was to have a viable vaccine within a decade. We are over three decades into the future, and safe, effective vaccines for malaria are still on the drawing board. Deaths due to malaria dropped by more than two-thirds using more prosaic methods — a remarkable achievement. The malaria parasite has proven to be much more wily than we initially anticipated. Also true of their brethren regardless of origin or ancestry.

Coda

We are unlikely to eliminate all human parasites, but we are morally obligated to attempt to control them as best we can. Many parasite populations, particularly those transmitted by food and water, can be reduced by improved hygiene. Clean water and better sanitation (1/4 to 1/3 of the world's inhabitants don't have access to toilets) would go a long way to solving some problems. We have excellent drugs for treating many of the intestinal forms. However, new drug development is non-existent mainly because the people affected are poor. There is no profit in the poor of the Third World. The Bill and Melinda Gates Foundation, the Carter Center, and others are doing heroic work to find solutions to these seemingly intractable problems with approaches at the high and low ends of the technological spectrum. If something works, use it.

Parasites are not going away. They are clever, resilient, frustrating, surprising, and, dare I say — beautiful. They are integral components of every ecosystem on Earth. If we were to remove them, all of them, we would undoubtedly have a different planet. I don't know if it would be better, worse, or just different.

4. The Beginning

You only live once, but if you do it right, once is enough.

— Mae West

Acowboy, fireman, or maybe the center fielder for the New York Yankees. I was pretty sure I would be one of those things, at least between the ages of six and 10. But a parasitologist? What kid wants to be a parasitologist? What kid even knows what a parasitologist is? I sure didn't.

I grew up in paradise. Paradise might conjure images of Hawaiian beaches, sunny southern California, or maybe the mountains of Colorado. I grew up in Youngstown, Ohio, now symbolic of the dying rustbelt, an area rife with unemployment, crime, and opioid addiction. My Youngstown wasn't those things in the 1950s and '60s. Paradise is as much about time as place.

On June 20th, 1949, I entered the world at 11:30 pm (a date and time which proved propitious later in life) at Northside Hospital. My parents, Ken and Jane Platt, lived on Almyra Avenue in half of a duplex they rented from my dad's oldest sister, Alice, and her husband, Ralph.

My father served in WWII from early 1942 until the end of the war, first as a shooting instructor and then in the Military Police. He had completed all but one course for a major in accounting from Ohio State University when he enlisted. As an accountant, he could have spent the war safely away from the perils of guns and bombs. Young men seek the camaraderie of violence, ignorant of the horrors awaiting them. My father joined the infantry. While he never saw combat, he did witness the aftermath of armed conflict during a short stint in the Medical Corps and the rubble of post-war Germany.

Like many soldiers returning from war, my dad wanted to get on with his life: earning a living and supporting his family. Following his service, Dad took the last course for his accounting major in night school at Youngstown College, later to become Youngstown State University. Another couple of credits in English Literature to complete his college degree were not critical. My mother graduated from high school at 16 and held several jobs; she was a whiz at shorthand and typing, until her husband returned from Europe. She then took on the role of homemaker, as did so many of her generation. My father worked for a small firm, McKay Machine, which made heavy machinery for the auto industry. Our family was solidly middle class unless we asked for something deemed extravagant (which covered most requests), then we were on our way to the poor house.

We moved twice before I left to attend college: from Almyra Avenue to Burma Drive before I formed any memories of my first home; and from Burma Drive to Macachee Drive when I was 11. Each move was two to three miles by car or bike, but the locations were so geographically and sociologically distinct they could have been different countries. Almyra Avenue was solidly working class: steelworkers, auto mechanics, and plumbers. Hardworking men doing their best to provide for their families. Burma Drive was the suburbs: office workers and small business owners. People who expected to achieve the American Dream and, for the most part, did. Macachee Drive was an escape from the monotony of the 'burbs; a few homes surrounded by woodland and streams with an eclectic mix of senior citizens, professionals, and blue-collar folks looking for something different.

My childhood was remarkably unremarkable. Nothing terrible happened to me. I had some broken bones and the usual childhood illnesses, but I was loved, fed, cared for, and educated by people I trusted. I was never abused or neglected in any way by family or acquaintances. I was in the vanguard of the "baby boom" generation. We were the center of attention, hardly "the seen but not heard" kids of bygone eras. My parents lived through the deprivations of the Great Depression and survived WWII.

They were told, I presume by Dr. Spock, to encourage us to express ourselves. Our parents weren't wealthy, but we didn't have to work outside of a few chores, mainly to build character. We were safe and allowed to roam the neighborhood and its surroundings almost at will. During the summer, we left home after breakfast, returned for lunch, left again until dinner, and again until it was time to go to bed.

We played baseball in the street with little fear of cars because our fathers took them to work, and few families had more than one. We played kick-the-can, rode bikes, and best of all, we were only a few minutes walk, or shorter bike ride, from Mill Creek Park, one of the larger municipal parks in the country. A quick trip down Burma Drive to Normandy Circle, across Bears Den Road, and you were there: teeter-totters, jungle gyms, swing sets, and baseball fields. During the summer, the city hired college kids to teach arts and crafts and organize ping pong and washer tournaments. In a park totaling over 2,500 acres, there were woods, streams, and caves to explore, rock formations to climb, and all manner of roads and trails to traverse on bike or foot. We were "free-range" kids before the term appeared in the lexicon.

Our home, 2247 Burma Drive, was a three-bedroom, two-story affair in a new suburb populated chiefly by young families similar to ours. Everyone had at least two kids, but three or four weren't unusual (I had two brothers). As a reflection from the perspective as an adult, I find it hard to understand how a developer, building a community for veterans of our most recent war, thought naming streets Burma, Oran, Normandy, and Coral Sea, all theaters of conflict in WWII, was a good idea. But it didn't matter. The houses were built, and veterans bought them.

The houses and lots were large for the time, but money was tight, and my father built the garage himself, hard to square with the man I knew growing up. I never learned anything about carpentry, plumbing, or auto repair (save how to change a tire) from him, although I did learn lessons far more critical: your word is your bond, be on time, and never look down on anyone.

From kindergarten to 5th grade, I attended Kirkmere Elementary, a 10-minute walk from our house. A few months before the 1960 presidential election, we moved out of the city to Boardman, an unincorporated township noted for having good schools and no African-Americans.

If race played a part in my parents' decision to move, it was never discussed when I was in earshot. West Boulevard Elementary School was my home for 6th grade.

Macachee Drive is located in the northwest corner of the township and close to invisible unless you lived there. The sign at the entrance read "Private Drive." A mile of single-lane blacktop through dense woods brought you to the first of nine homes. There had been a lake behind our house, a big reason my mother loved the property; however, we never saw it full. The lake became derelict with a stream and secondary growth, forming a vast playground to explore. I collected fossils, mostly crinoid stems and brachiopod shells, and all manner of wildlife.

I was fascinated with science and nature for as long as I can remember. When I was eight or nine, I begged to stay up until 9 pm to watch the Bell Laboratory Science specials with Dr. Research. The episode I recall most vividly was about the sun and featured a segment on photosynthesis. An animated chlorophyll molecule described the process of converting light into chemical energy but hid the specifics behind a screen and told Dr. Research to "figure it out for himself." The details of the process were still a mystery.

I was never squeamish about blood. As a child, I bit my nails ferociously. On one occasion, two of my fingers became painfully swollen and required lancing to relieve the pressure and clean out the infection. Dr. Short, a local physician, numbed the digits with Novocain and told me to turn my head while he operated. I said I wanted to watch. Dr. Short looked at my mother. She shrugged her shoulders, silently affirming my decision. I watched in fascination as the blade pierced my skin. I didn't feel a thing! Dr. Short drained the pus, treated the wounds with antibiotics, and bandaged them. He gave my mother a prescription for oral antibiotics, and I got a stern lecture to stop my "nasty habit." I was mesmerized. I didn't stop biting my nails; however, I was more precise.

My first experience with parasites occurred at the age of five or six. My mother diagnosed and treated me for pinworms. Yes, we were middle class, and yes, I had a pediatrician — Dr. Harold Segal. Nice people, however, didn't get pinworms, and we were nice people. This was something Mom thought the doctor didn't need to know. My mother's treatment was a series of salt-water enemas until no worms appeared in the toilet. I do recall being encouraged to look in the bowl and see the small, white worms slithering across the surface of a floating turd. I don't know what I thought — if anything. I know I didn't think — hey, I want to be a parasitologist!

Nice people do get pinworms. *Enterobius vermicularis* is common in children and an occupational hazard for anyone who has extensive contact with children: elementary school teachers, daycare workers, and parents. It is one of the few parasitic infections not associated with poverty. I diagnosed pinworm infections in children of friends and colleagues over the years. I tried to assuage their feelings of inadequacy and guilt, assuring them they were nice people. I explained there were excellent treatments available, and they should take their child to their physician. Dr. Segal never learned of my infection or my mother's dive into the practice of medicine.

The best day of my childhood revolved around baseball, not the game or a game, but the trip to a game. Once a year, Uncle Tom (my mother's brother) took my cousin Doug and me to Cleveland to see the Yankees and Indians play. I was a Yankee fan for no other reason than I asked my older brother, Ken, who was the best player in baseball. I was five or six at the time. His response? — Mickey Mantle. I asked who he played for, and the answer was the New York Yankees. They were my team. Doug, a Buckeye all the way, cheered for the Cleveland Indians. The Yankees almost always won, and I was insufferable.

In 1959, we were going to a game in mid-summer. I was 10, and Doug was 12. Passenger trains still ran regularly between the two cities, and we were downtown to board for the 60-mile trip to Lake Front Stadium. Passengers crowded the train, and there weren't three adjacent seats. Uncle Tom didn't want the two of us sitting together without adult supervision or one of us sitting alone. The conductor noticed his dilemma and offered a solution beyond our wildest dreams. Would we like to ride in the mail car? Our response was an immediate and enthusiastic — Yes! Doug and I lounged on mail sacks and viewed the passing countryside through an open door on our way north, a real Tom Sawyer/Huck Finn moment. The Yankees won, of course.

My path through high school was in the seams. I was not a jock even though I was on the swimming team at the YMCA. Boardman High did not have a swim team, so I was not an athlete in the eyes of my classmates.

My grades were OK: A's, B's, and a few C's. I wasn't a brain and never hung with the smart kids. I didn't get in trouble, so I wasn't a hood. And my family wasn't in the right social set, nor did we live in the more upscale part of the community to be part of the popular class. I knew and got along with kids in all of the cliques but was not part of any of them. High school was neither a good time, although I have some fond memories, or a bad time. I wasn't bullied or picked on. It was a time to be endured. I stayed in the Boardman school system until I graduated in 1967. Moving to Boardman was difficult. I went from the leader of our gang of adolescents at Kirkmere Elementary to nobody in Boardman. I stayed a nobody until I left for college in 1967.

The only class in high school I remember with any fondness was freshman biology. Our teacher, Alan Burns, was young and a star basketball player during his high school years at Boardman. I don't remember any specifics of the class other than Mr. Burns was energetic, enthusiastic, and made the material come alive. My fondness for exploring the natural world, first in Mill Creek Park and later in the woods around our home on Macachee Drive, probably made me more receptive to the study of biology, particularly with a young, charismatic teacher. Mr. Burns' enthusiasm for biology reinforced my desire to pursue science in college.

While I never had anything terrible happen to me, one incident profoundly affected our family. When my younger brother Bill was five, and I was 15, my mother organized a playdate for Bill with the daughter of friends from the Unitarian Church: Gordon and Diane Brott. Mom was going to take Allison and Bill to see *Mary Poppins* at one of the downtown theaters. I was going downtown to the YMCA and in need of transportation. Diane was late delivering Allison, and I was an impatient jerk. Even though the 'Y' was seven miles away, I told my mother I wasn't waiting any longer and headed out on my own, walking and hitchhiking to my destination. I spent the afternoon swimming and playing ping pong. Late in the afternoon, I hopped on a bus for my aunt's house on Almyra Avenue; buses didn't get to Macachee Drive. I would call my parents to come and pick me up.

When I arrived at my aunt's, I was greeted with near panic by my relatives. They had been trying to track me down all afternoon. My mother had been in an auto accident, and she and Bill were seriously injured. I was quickly bundled into a car and taken to the hospital and found my father slumped in a chair in the waiting room. The only time I recall seeing my dad other than the reliable, confident provider typical of men of his era was when one of us (mostly me) was hurt. My mother and brother were both in surgery. The prognosis was good, but dad was powerless; there was nothing he could do but wait and trust the surgeons' skill. Allison, the girl my mother had been waiting for, was dead.

As the hours and days passed, a picture emerged of what had happened that afternoon. My mother drove a Volkswagen beetle. When Allison's parents arrived, she and my brother were bundled into the back seat; Bill behind my mother and Allison behind the empty front passenger's seat I would have occupied. As they drove down Glenwood Avenue toward the theater, a car traveling in the opposite direction crossed the centerline, headed directly at my mother. Mom jerked the wheel to the left in an attempt to avoid the crash. The oncoming vehicle hit the right side of the bug crushing the passenger compartment. In an era before locking seats and safety belts, Allison was hurtled forward and crashed through the windshield. She died instantly.

A few days later, with my mother and brother out of danger, my older brother, Ken, drove me by the salvage yard where the police towed my mother's crushed beetle. One look conveyed an awful truth. I should have died that day, and Allison might well have survived. A fit of pique saved my life. If I stayed, things might have been different. Perhaps we would have gotten underway a minute sooner or a minute later and avoided the crash entirely. Any song from Mary Poppins reduced my mother to tears. We never owned another Volkswagen.

I planned to go to college; however, there were times my father wasn't so sure. After one mediocre report card, he threatened to send me to carpenter's school if I couldn't do better. I only applied to two schools: Hiram and Miami University (Ohio). I never seriously thought about Miami. Education did not drive my decision to attend Hiram. A friend of mine and former teammate on the YMCA swim team, Russ Cargo, was attending Hiram, a small liberal arts school about 40 miles north of Youngstown. I wanted to swim in college, but I wasn't an outstanding athlete. However, Hiram's swimming team was mediocre at best. It was a match made in heaven.

I entered an age group swim meet held at the college and met the coach, Bill Donaldson. Coach Donaldson was a bear of a man with a smile almost painted on his face. He liked everyone, and everyone liked him. After a race, we talked briefly. He flattered me, and I was hooked. Hiram did have an excellent academic reputation and a well-regarded biology program. My parents were happy, and I pretended to care.

My senior year in high school completed, I worked at Youngstown Cartage, a trucking firm specializing in hauling steel. They had a warehouse where they stored material for later shipment and employed half a dozen mechanics to maintain their fleet of trucks. As a warehouse laborer, I got every shit job they could find. I unloaded large bales of wire from railroad cars and emptied a coal bin by hand to get to the broken screw at the bottom that fed coal into the furnace. I cleaned windows caked with diesel fumes using an acid solution that ate holes in my clothes. And, of course, I loaded and unloaded flatbeds with rolls of steel weighing 30–40 thousand pounds.

Safety measures were lax, and I was ever vigilant to avoid injury or worse. The crane operator nearly killed or maimed me more than once. One morning we were tasked with loading a massive steel mill component onto a flatbed. The piece was top-shaped with the point cut off, and we made a cradle out of chains to lift it. Several of us pressed the chains flat against the sides of the "top" with the palms of our hands while the crane operator slowly took up the slack in the cables and tightened the cradle on this oddly shaped and massive piece of metal. The cables rose, and I felt the chains tighten as I pressed them against the surface with my fingers outstretched.

Unexpectedly, the crane operator reversed direction, and the chains began to drop. Instinctively I closed my hands to keep the chains on my side from falling. In a split second, the operator jerked the cables to raise the load, and I felt the chains moving up toward the mass of metal. I threw my arms down to my sides and watched the "top" rise in the air with both of my work gloves pinned between the chains and the steel. If I hadn't responded reflexively, I would have lost all four fingers on both hands, leaving only my thumbs as a reminder of my co-worker's incompetence.

Youngstown Cartage wasn't all bad. I had a great relationship with the diesel mechanics. They worked hard, loved practical jokes, and, while not book-smart, had more common sense than some PhDs I've met in my life. They had a great time making fun of the "college" boy, although I was still months away from enrolling. At some point during the summer, each of them took me aside and told me, with a level of seriousness I never witnessed at any other time, "Work hard in school; you don't want to do this for the rest of your life." I wish I could say I took their admonitions to heart sooner rather than later. I finished the summer and was more than delighted to collect my last paycheck. I couldn't wait to start school, at least the swimming part.

5. Hiram College

Teachers open the door, but you must enter by yourself.

— Chinese Proverb

Freshman Year: 1967–68

Hiram, founded by the Disciples of Christ, was one of many liberal arts colleges established by religious denominations in the Midwest during the 1800s. My freshman year students numbered about 1,200, and the town was home to 250 residents, mostly college faculty and staff. The village boasted a small grocery store, a post office, and one stoplight. The nearest town of any size was Garrettsville, about three miles east, with a population of maybe 1,500. The college, like most at the time, was conservative and functioned in *loco parentis*. Almost all students lived in the dorms, dinner served family-style, and dress for Sunday brunch was coats and ties for the men, dresses (or something similar) for the women. The campus was dry, which meant no alcohol in the dorms. Women had parietals and had to be locked in no later than 11 pm on weekdays and midnight on weekends.

The Biology Department had an excellent reputation; why, I had no idea. There were only four faculty. Dwight Berg taught all the botany classes. Jack Miller taught anatomy and genetics. Bill Cool, a PhD student at nearby Kent State University, handled ecology and physiology. Zoology was the domain of Jim Barrow. I never had a course with Dr. Miller. Bill Cool was a non-entity. He did his job and disappeared to work on his doctoral research, a concept utterly foreign to me. Berg and Barrow were the driving forces in the department and couldn't stand each other. They were of the same generation but with diametrically opposed personalities.

Dr. Berg was formal, intelligent, and seemed to have been born old. He loved plants, and it came through in his lectures even though the subject was of little interest to most students, even the biology majors. His compulsive refrain of "Do you follow me?" was close to metronomic. I counted over a dozen repeats in a single 50-minute lecture. What struck me was the enthusiasm Dr. Berg conveyed during class. My assessment of his presentations was, "I found out something utterly fascinating, and I want to share it with you." His enthusiasm was undiminished even though he taught the course three times a year for 17 years when I occupied a seat in his class. I realized much later in my academic life that it didn't matter Dr. Berg was repeating much the same material year after year; it was new to us. He did everything in his power to get us excited. It should have been infectious, but he was more hypnotic than captivating. I took three classes from Dr. Berg and earned C's in all of them.

Jim Barrow could have been a character in the movie *Midnight in the Garden of Good and Evil*. He was a good old boy (Georgia, as I recall) through and through. With his southern drawl and hair combed back reaching his collar, Dr. Barrow charmed parents into trusting him with their children and their children into doing almost anything he asked. He never thought small, but I sensed he was a lazy academic. I don't recall him ever using notes in the many classes I took from him during my four years. My assessment, again a later reflection, was that he related what he memorized during his years at Yale and did little to stay current.

Dr. Barrow conceived, started, and ran the "Station." He convinced benefactors of the college to purchase a local farm, about three miles from campus, for the college to use as a site for field studies in biology. He charmed and cajoled students into doing the work necessary to convert the place into a biological field station. It consisted of farm fields gone fallow, woodlands, a small pond, and a few remaining outbuildings typical of a working farm. If you majored in biology, Dr. Barrow expected you to spend time doing whatever project he had ongoing at the time. The more hours you spent, the higher your position as one of Jim's favorites.

I spent more time at the swimming pool than in the library or the field station, and it showed during my freshman year. My times improved substantially, and I was named MVP, the first of three such awards during my four years. I could memorize swimming times at a glance, a rarity in my academic pursuits. Coach Donaldson told us at our first meeting as a team each year: "You can do three things here: academics, athletics, and party. You can do two of them successfully, but not all three." He was right.

My entry into academic life was far from smooth. By the time I registered for classes (first-year students were last), none of the courses I wanted were available. My first quarter consisted of Calculus I, Introduction to Botany, and Introduction to the Theater. Hiram had a unique quarter system: three courses per term, each class met four days/week, with no classes on Wednesday. We loved having Wednesday off. Serious students used it to study and catch up on assignments, while the rest of us had an extra day to party.

Wendell Johnson tried valiantly to teach me calculus. I was never sure what Keith Leonard was doing in theater, but I learned to love live theater in spite of him. I was fascinated by Dwight Berg, but botany held little interest. I focused on swimming and partying and scraped through my first term with straight C's. The end of our freshman year culminated with the assassinations of Dr. Martin Luther King in April and Bobby Kennedy in June. Coupled with the escalation of fighting in Vietnam, student unrest was evident even in the sleepy, rural environs of Hiram, Ohio. Jobs were scarce, and I returned to Youngstown Cartage for a second and final summer trying to avoid being maimed or killed for a buck 75 an hour.

Sophomore Year: 1968–69

The summer of 1968 brought radical change to Hiram. As we began our sophomore year, home-style meals were gone, as were coats and ties for Sunday brunch. We could drink beer in the dorms, and men and women could visit each other in their rooms until midnight on weekdays and 2 am on weekends. A new student center replaced the previous structure, an old Quonset hut built during WWII. The hippie movement hit campus, and boys with long hair and bell-bottoms were the norm, as were girls in miniskirts. Unfortunately, those changes, while welcomed by the students, did little to stimulate my academic performance. I was still committed to majoring in Biology, but my thoughts of attending medical school (What else did you do with a degree in science?) were rapidly fading.

Racial tensions rose across the country in the wake of Dr. King's assassination. Rioting broke out in several major cities during the summer. Hiram had a small contingent of African-American students, but none on the faculty or in the administration. Protests started slowly on campus as some African-American students petitioned for courses in African studies, black faculty, and advisors. Tensions came to a head about halfway through fall quarter. A small cadre of black students barricaded themselves inside Hinsdale Hall, the main classroom building, and issued a series of demands. Curious students gathered on the lawn and listened as the protesters broadcast their grievances from a second-floor window.

Initially, there was a sense of amusement among the predominantly white student body, but things turned ugly as the day wore on. A few guys, dressed in sheets mimicking the robes of the KKK, circled the building shouting racial epithets. As night fell, rumors circulated suggesting the Cleveland chapter of the Black Panthers was sending some of their members to campus in support of their "brothers and sisters." Black Panthers did not come from Cleveland or anywhere else. The incident ended peacefully when the administration met with the protest leaders and promised to address their issues. The college made changes, and no further overt clashes over racial issues occurred, but some personal relationships were strained or irrevocably broken.

I was introduced to the concept of scientific research by none other than Dwight Berg. I took Introduction to Botany in my freshman year and exited with a gentleman's C. If I was going to major in Biology, I had to take more courses from the man. First-quarter, I took Non-Vascular Plants, a survey of fungi, mosses, liverworts, and other organisms lacking the vascular tissue (xylem and phloem) found in higher plants. Each student was required to make a collection of one of those groups of organisms, identifying each specimen to species and categorizing them according to the current classification scheme. I chose fungi. By the end of the quarter, I had two shoeboxes containing nearly 100 mushrooms and related specimens.

Fungi were my introduction to biological systematics — taxonomy, nomenclature, and classification. I enjoyed collecting in the wooded areas surrounding the campus, which harkened to my days exploring Mill Creek Park and the woods around our home on Macachee Drive. I recollect earning a good grade, a B or B+ for the collection, but my performance on exams resulted in another C from Dr. Berg. Little did I realize a good deal of my working life would focus on those aspects of biology, but studying parasitic worms — not Dr. Berg's beloved plants.

I enrolled in Vascular Plants in the spring and another tussle with Dr. Berg. That semester's fare was the flowering plants and their relatives containing xylem and phloem. We studied the anatomy, morphology, life cycles, and evolution of these familiar and ubiquitous organisms. Laboratories included a comparative examination of various groups of plants and field identification. Each week we hiked through the forests and grasslands near campus or visited nearby parks harboring interesting or rare species not found close to school. Dr. Berg was a dynamo. We had to jog to keep up with him while he chain-smoked Lucky Strikes. If you fell behind, you could always find the head of the line by the thin trail of smoke emanating from the good professor. How he survived into his 90s, I will never know.

Near the end of the term, Dr. Berg organized an afternoon excursion around campus as a review for the field exam scheduled for the end of the week. At one point, I noticed a plant I didn't recognize. I pulled it up and asked Dr. Berg what it was. He took one glance, shook his head in dismay, and said, "poison ivy." I sank even lower in his estimation.

As far as I have been able to determine, Dr. Berg never published any research after completing his PhD. However, he placed a high priority on student research in his courses. Presented with a list of possible projects, I chose "Sporangium development in three species of the fern genus *Osmunda.*" Sporangia form in clusters called sori on the underside of the leaves of the sporophyte, the stage we recognize as a fern. Each sporangium contains spores that possess a complete set of chromosomes. Spores germinate, producing a gametophyte, an inconspicuous part of the fern life cycle, which produces eggs and sperm — fertilization results in the production of the leafy sporophyte. I collected the sporophyte leaves, cut out the sori, and embedded them in paraffin, a wax-like material that stabilized the tissue for sectioning. Sectioning involves mounting a specimen in a block of paraffin on a stub, inserting the stub into a microtome, a device for slicing the tissue into less than paper-thin sections, collecting the sections on microscope slides, and staining them. The stained specimens were then examined microscopically. I loved this part of the class and produced a paper much better than my exam grades. Nevertheless, I earned my third and final C from Dr. Berg.

I escaped the purgatory of Youngstown Cartage and was hired for the summer as the Head Life Guard and Swimming Coach at the Boardman Swim Club. I was in heaven.

Junior Year: 1969–70

Junior year proved to be a turning point from a variety of perspectives. My first two years centered on swimming and partying. My academic performance showed it with a GPA hovering around 2.5/4.0. I finally decided Coach Donaldson was right, and if I wanted to do something with my life partying would have to take a backseat to studying and swimming. I realized I could do the academic work and enjoyed it. My GPA in my last two years was over 3.5. Despite my improved academic performance, I knew I wasn't going to be a doctor or dentist. In the absence of those options, I began to think graduate school might be a possibility.

I spent almost all of my time in the science building — Coulton-Turner Hall. I was taking courses in Zoology, which interested me much more than botany. I fell under the charismatic sway of Jim Barrow and became one of Barrow's Boys. I spent time at the Station doing grunt work and became peripherally involved in a research project on the territorial behavior of Canada geese.

One of my favorite classes was Invertebrate Biology, taught by, of course, Jim Barrow. He arranged a trip for the class to the University of Delaware's marine station in Lewes. We collected marine organisms for identification and observed the behavior of others in beach, marsh, and littoral habitats. Dr. Barrow provided my first trip to the coast and my first experience with an ocean. We ate at a local seafood restaurant, and I had raw oysters for the first time. Even though it was early November and the temperature hovered in the low 50s, the Gulf Stream brought warm southern water to the eastern seaboard, and most of us decided to take a dip in the Atlantic. It was a glorious four days.

I was finally succeeding academically, and my efforts in the pool were still yielding results. While these achievements were significant to me for a whole host of reasons, my parents' approval most notably, they weren't the most important events of the year. Three things were more consequential, and two would play a significant role in shaping the rest of my life.

On September 27th, 1969, I was in the student union watching two friends, Wes Ogata and Steve Mawby, play pool. Wes was on the swim team, Steve was a football player of some renown, and we were all in the same fraternity — Phi Gamma Epsilon. A beautiful blonde appeared out of nowhere and began talking to Steve and Wes. I knew who she was. It was a small school, and you had to work hard to remain invisible. Her name? Kathy Fenton.

We met briefly at a dance during Orientation Weekend freshman year. I only saw Kathy on a couple of occasions during the intervening years. We had no classes together and hung out in different circles. Besides, she had a reputation as being a prude, and I was hoping for something else.

Kathy carried most of the conversation. I joined in occasionally because she was (and still is) cute, and I was interested. At some point, the conversation turned to alcohol. Kathy admitted she didn't like to drink but thought she might like bourbon and ginger ale. I saw my opening. I suggested we go back to my room for a drink. I didn't have any bourbon, but I figured I could scare some up without too much difficulty. Yes, we were underage, but alcohol was never difficult to find on any college campus even then. Kathy was taken aback by my rather brazen overture and claimed she couldn't because, as the RA in charge of Henry Hall, she was responsible for locking up later in the evening. Undeterred, I pressed her for a time we could get together and share a bottle of bourbon. She finally relented, and we agreed to our first date, the following Saturday — October 4th, 1969. I didn't know it at the time, but I had a date with my future wife.

Since I promised bourbon, I made a trip to the liquor store in nearby Mantua and returned to campus with a bottle of Jack Daniels. Our first date would be a threesome: me, Kathy, and Jack. The following Wednesday, a friend and member of the swim team, Herman Kling, informed me my date with Ms. Fenton might not happen. Herman talked with Kathy during lunch and mentioned our upcoming rendezvous. Kathy told Herman she didn't think I was serious and planned to go home for the weekend. When

I heard this, I got on the phone immediately to confirm my intentions were serious. Kathy was gracious (as always), and the date was back on track. Who knows what might have happened if the conversation hadn't taken place, if the message hadn't been transmitted in a timely fashion, or not at all?

I decided I wanted to make it a memorable evening, so I enlisted the aid of one of the other residents on the floor. Butch Weir, a senior, had a Triumph Spitfire, and I convinced him to be my chauffeur for the roundtrip from the quad to Henry.

At five to eight, Butch and I got into his car and drove the two blocks to Henry. As my "man" for the evening, Butch approached the front desk while I took a seat in the lounge. I heard him ask the student on duty to inform Ms. Fenton that Mr. Platt was there to pick her up for their date. I can't imagine what the desk worker thought, but she placed the call. When Kathy came down from the 2nd floor, she was a bit bewildered when greeted by Butch, adorned in a raccoon coat, a pork pie hat, and sunglasses. Butch walked her over to where I was sitting and made the appropriate introductions. My man led the way out of the dorm, opening doors for us as we exited. When we got to the car Kathy was a bit perplexed; first, it was maybe a three-minute walk from Henry to Bancroft, and second, there were three of us and only two seats in the Triumph. Since Butch was driving, the only solution to the problem was for Kathy to sit on my lap. The plan worked wonderfully!

Before leaving for Henry, I gave Butch my room key. When we arrived back at the quad, Butch preceded us, opening doors as we climbed the stairs to 3rd Gray and through the swinging doors separating Gray and Bancroft. When we arrived at 321, Butch had my key in hand, opened the door, and asked: "Will there be anything else?" I indicated his services would not be needed, and he could have the rest of the evening off. Kathy was mystified and amused by the little comedy I planned to begin our evening, but I think she was flattered by the preparation and attention.

The principal players were in place (don't forget Jack!), and our evening together was underway. For the first hour or so, we listened to music, watched a little TV, and talked. Finally, Kathy asked for a drink. I got a couple of beer mugs, ginger ale, ice (we had a fridge), and the bottle of Jack

Daniels. We both drank enough to get a bit tipsy (she did like bourbon), but not so much as to do something she might regret. I walked her back to Henry (remember, Butch had the rest of the night off) around 1 am.

We became a couple. It would be nice to say we continued as a couple, got engaged, and married with no hiccups. Over the next four years, we broke up four times — two for me and two for her. There was chemistry; we just had a hard time balancing the charges.

The second life-changing event happened on December 1st — the first Draft Lottery. In a push to get more soldiers for Vietnam and make defending our country less dependent on minorities and the economically disadvantaged, the government decided to stop giving student deferments for military service. They would honor those already granted, but upon graduation (or flunking out), the young men who had escaped conscription were classified as 19-year-olds with 1A status. Each year's crop of 19-year-olds would be drafted according to their birth dates.

The order would be determined by lottery, pulling plastic capsules out of a drum until all 366 days of the year were accounted for. The order was supposed to be random (it wasn't). Young men gathered in bars, dorms, student centers, apartments, and living rooms to see their fate decided on television.

I had swimming practice from 4–6 pm and arrived at the dorm after the first 50 (approximately) numbers were out. The mood in the room ranged from rage and despair (already chosen) to hopeful if their birthday was still tumbling in the wire-mesh drum. The rumor preceding the "big gamble" was that no draft board would go beyond 195 to fill their quota. A low number meant military service: most likely Vietnam following graduation. A high number meant safety unless you were crazy enough to enlist. I didn't share a birthday with anyone in the room, so nobody could remember if my birthday, June 20th, had been pulled. I sat in limbo, watching as more and more dates were announced, not knowing whether I was safe or my future was already determined. By the time they reached 350, I resigned myself to a stint in the military. But, miracles of miracles, with only seven capsules left in the drum, they pulled number 360 — June 20th! I was out. I was not going to Vietnam. I could honestly think about what I might do after graduation and not slogging through the jungles of Southeast Asia fighting a war I did not support.

Within a week, I wrote my draft board and requested reclassification from 2S (student deferment) to 1A (available for service). I was betting the rumors were true, and my draft board would not need to go past 195 to fill their quota. I was, thankfully, correct. Years later, looking at the complete list of dates and numbers, I realized how close I came to being draft bait. Recall I was born at 11:30 pm on June 20th. Another 30 minutes in the womb, and my birthday would have been June 21st — number 60. A number low enough to virtually guarantee an induction notice.

The final episode in the Trinity did not change my life but touched too close to home with tragic consequences for a high school classmate and friend: the National Guard Massacre at Kent State on May 4th, 1970. Kent State University is approximately 20 miles southwest of Hiram. Nixon's expansion of the war fomented student unrest and protests on campuses around the county. Kent was a hotbed of student activism. We had a small cadre of anti-war types at Hiram. They made a little noise but nothing bordering on anarchy. I was not among them. My older brother, Ken, was a lieutenant in the Navy stationed somewhere in the Mekong Delta patrolling the river and ferrying Navy Seals to and from missions. While our mother adamantly opposed the war, I couldn't be disloyal to my brother. I opposed our involvement, but I couldn't bring myself to protest actively.

On May 4th, news of the shootings swept through our campus like a tsunami. We had to wait for the evening news to get accurate information about what happened. While watching Walter Cronkite, I learned Sandy Scheuer, a high school classmate, was one of the four students killed. She was not part of the demonstration but was walking to class approximately 150 yards away when a bullet struck her in the throat. She died where she fell. I was sick. Nobody could believe soldiers would open fire with live ammunition on a bunch of college kids.

My junior year was one of the most tumultuous in memory. Escalation of the Vietnam War unleashed campus unrest across the country. Anti-war protests at Hiram paled in comparison to those at other schools. I avoided the political turmoil and focused, for the first time, on academics, which improved dramatically — and swimming. I qualified for the finals in two events in our conference championships and was named MVP for the second time in three years. As the end of college life began to come into focus, what I might do following graduation had not. There were only two things I loved: swimming and biology. I weighed the possibility of career paths in each: coaching or graduate school.

I returned to my role as the head lifeguard and swimming coach at the Boardman Swim Club. I supervised a staff of six guards and coached about 100 competitors ranging in age from 6–18 through the summer league competitions. I typically arrived before 8 am to get ready for practice and frequently didn't leave until the pool closed at 10 pm. Despite the long hours and responsibilities, I enjoyed the job. The swimmers, especially the young girls, and parents loved me. The pay wasn't great, but working 80–100 hours a week, with little time to spend any of it, resulted in a reasonably healthy bank account by the end of the summer. I also realized coaching was probably not what I wanted to do long-term. It was great for a summer, but not a lifetime.

Senior Year: 1970–71

Senior year was perhaps the least eventful of my four at Hiram. Kathy graduated the previous spring and was teaching Physical Education at a middle school in Columbus, Ohio. I had a few dates during the year, but nothing of consequence. I was fully committed to my studies. I loved the biology classes I needed to complete my major and some electives just for fun. My love of swimming, at least the workouts, was waning. My times didn't improve at the same rate as in previous years but came down in niggling increments. I was awarded the MVP trophy for the third time in four years. We compiled a record of 10–3, the best season in Hiram's history. My swimming career was over, but not my love of the sport.

What are you going to do when you graduate? The question wasn't as fraught in the '60s and '70s as it is now. If asked, most of my friends shrugged their shoulders and replied, "I dunno, get a job?" The more I pondered the question, the more I realized 1) a career in medicine was not in the offing; 2) I didn't know anything anyone in their right mind would pay me for, and 3) graduate school was looking like the most promising alternative. I talked with Jim Barrow about the application process and got started. I completed applications for ten universities across the country. I don't recall the criteria I used, but I stuck to middle-tier state schools because my overall GPA hovered around 3.0/4.0. I wasn't shooting for Harvard or Berkeley.

Since I was applying to Master's programs in Biology, almost every form had a space to identify a field of interest. Every time I finished a class (except botany), it was my favorite. I was taking Parasitology when I filled out the applications, so I wrote Parasitology. I figured if I got there and didn't like it, I could switch to something else. It was another one of those choices, trivial at the time, but in the immortal words of Robert Frost, "I took the one less traveled by. And that has made all the difference."

During spring quarter, I was cruising through the final classes of my last term at Hiram. Swimming was over. Letters from graduate programs began appearing in my mailbox. Most were rejections. While my grades for junior and senior years were mostly A's, my less than stellar performance as a freshman and sophomore wasn't attractive to most universities. Three schools, New Mexico State, West Virginia University, and Bowling Green State University (OH), accepted me. BGSU was the only one to offer a full graduate fellowship. They bought me for $190 a month and a tuition waiver in return for teaching (which I had never done) biology laboratories. I also received a letter from Dr. Francis Rabalais. As the parasitologist on the faculty, he would be my advisor and suggested we meet in person to discuss his program.

A few weeks before graduation, I borrowed my parents' 1966 Dodge Dart and drove west on US 224 for Bowling Green. Dr. Rabalais' office was on the 3rd floor, and he was at his desk when I arrived. In the pre-internet/Google era, I had no idea what to expect.

Rab, as he was known to his students and colleagues, was short and stocky with a shock of curly red hair. His sideburns were long, with a mustache and goatee surrounding a snaggle-toothed grin. He welcomed me warmly and got down to business. He would not be my advisor; he would be my major professor. He wasn't there to make *suggestions* about my education; he would educate me. And he couldn't do his job if I wasn't in the lab. He expected his students to be at their desks no later than 8 am. Everyone went home around 4 pm, ate, took a nap, returned no later than 7 pm, and stayed until midnight. Students could take off either Friday or Saturday evening, but not both. Rab maintained the same schedule, except his day ended at 10 pm. Rab's expectations were high. He permitted one B in your career. If you got more than one, you had to find a different advisor with lower standards. Take it or leave it. There was no negotiation. Rab was the house, and those were the rules. I wasn't sure what I was getting into, but I had two choices: accept or do something else. We shook hands, and I left to look for a place to live in the fall. The few vacancies I found were dismal, and I opted for a graduate dorm until I could get my bearings and find something suitable.

My final weeks at Hiram were idyllic. I didn't have to worry about final exams as the college had an unusual and welcome tradition. Third-quarter seniors could opt out of finals and accept the grade they had at mid-term as their course grade. Mine were excellent, so I didn't have to prepare for finals. I rarely cut class and completed all my assignments. I attended classes because I enjoyed them. I loved "Science in 19th and 20th Century British and American Literature" taught by a husband and wife duo, a physicist and literature professor. We examined Charles Darwin's influence on Thomas Hardy, Albert Einstein on Lawrence Durrell, and many others. It was great fun and my first real exposure to the interrelated nature of different disciplines.

I had one final encounter with Dr. Berg. I chanced upon a group of students, mostly seniors, in one of the laboratories talking with Dr. Berg and joined them. It was one of the few times I recalled seeing him relaxed and not the faculty member perennially disappointed in my classroom performance. The conversation ranged over various topics but finally got around to "What are you doing next year?" When it was my turn, I shared I was going to graduate school. Dr. Berg stopped and stared at me in utter disbelief. Remember, I earned, and I did *earn* them, C's in every class I took with him. He didn't say a word, but I can only imagine him thinking — "In what, English Literature?" — because it couldn't be biology.

I attended college from 1967–1971, the era of sex, drugs, and rock 'n roll. I could only claim familiarity with rock 'n roll. I did not, then, or ever, use illegal drugs. Bill Clinton claimed to have smoked marijuana but never inhaled. I claim the reverse. I never smoked, but I attended parties where others did — in profusion. It was impossible not to inhale. As far as the sexual revolution was concerned, I was a non-combatant.

As a noteworthy aside, in 2018, Hiram named me as the recipient of the J.J. Turner Award for Excellence in the Life Sciences. Dr. Dwight Berg was an early inductee, and died a few years before I received the award. I can only imagine him spinning in his grave if such things were possible. Later that year, I was inducted into the college's Athletic Hall of Fame. I took one botany course in graduate school. My grade? An A!

I graduated, packed my things in my parents' car, and headed back to Youngstown. I wasn't there long as my older brother, Ken, invited me to spend the summer with him in San Diego, where he was stationed in the Navy. My parents gave me a round-trip plane ticket as a graduation present, and I was on my way to sunny southern California. My time in the west was unmemorable. I had little to do but ride a bike to the beach, read, and do some housework in Ken's apartment. He was on duty and had to be at the naval base every day.

In late July, I received a letter from my mother containing a disturbing note from BGSU. The demand for graduate student housing was too small to warrant accommodating those who had requested it, so we were on our own. I didn't have a place to live, and classes started in early September. I had to get home to find an apartment. I made a spectacularly idiotic decision. I cashed in my return plane ticket and prepared to hitchhike the 2,500 miles back to Youngstown. I packed my suitcases and put them on a Greyhound for delivery to Youngstown, loaded a backpack with a few essentials, and Ken dropped me off at the onramp to the interstate. I don't recall Ken trying to talk me out of this insane odyssey. If he did, I ignored him. My parents were unaware of my travel plans and how little common sense four years of college tuition buys.

My first ride was with a guy who sheared sheep for a living and drove a pickup truck held together with wire and duct tape. He left me in El Centro. El Centro is in the middle of the desert, with temperatures routinely in triple digits. I walked to the Holiday Inn near the onramp to get some water, and the front desk clerk shared he had seen single women out there for three days with no luck. Despite this dismal prospect, I walked back out to try to catch a ride. A few minutes later, a couple picked me up and screamed at each other for about an hour before dropping me off in Yuma, Arizona. It wasn't any cooler, so I went into a bar for the air-conditioning and some lunch. I must have been severely dehydrated and downed three glasses of water before my meatball sandwich arrived. After finishing, I went back out to the highway in search of transportation and unceremoniously evacuated my stomach of everything I had eaten.

I landed a ride with a guy recently returned from Vietnam who took me across Arizona, New Mexico, and all the way to Dallas. I spent the night in a motel and started early the following day heading north and east. A kid dressed in a tee-shirt and boxers took me about three miles (he wanted to be helpful), and a businessman of some sort got me 100 miles closer to my destination. At this point, I was in the Middle of Nowhere, Texas, as the interstate system was far from complete in 1971.

A pickup truck stopped, and I hopped in. I was with a real-life rodeo cowboy. He was going home to see a girl he dumped two years earlier and was hoping to rekindle the flame. After a mile or so, he made a U-turn, drove back past where he picked me up, and turned down a dirt road into the woods. He stopped in front of what appeared to be an abandoned building, jumped out of the truck, and jogged inside. I had no idea what he was doing. Should I get out and run for it? Before I could gather my wits and do anything, he came bounding back out, holding four cans of beer. He got in the truck and headed back toward his destination with the cans tucked between his legs. He popped the top off one, took a swig, and asked: "Want one?" Those were the first words out of his mouth since before the U-turn. I declined with the excuse 10 in the morning was a little early for me. After 20 miles, he pulled onto the berm and said he was turning, and I got out. He wished me well, and I told him I hoped things worked out with his girlfriend. There was only one beer left unopened. I would have loved to have been a fly on the wall for that reunion.

I was near Texarkana, Texas, and over halfway home after less than three days on the road. I quickly got a ride to Nashville and decided to take a bus to Columbus. Kathy lived there, and like the cowboy, I harbored the fantasy of a warm reception. I called her and learned she was visiting her folks in Louisville (OH) for the weekend. One of her roommates, a friend from Hiram, offered to pick me up. After a quick lunch, I decided there was no point in staying in Kathy's absence. Her roommate drove me to the interstate, and I hopped out at the onramp to I71 north. My luck held, and I got a ride within two miles of my parents' house. Why a single woman with a toddler in the backseat would pick up a single guy, I have no idea, but I was grateful. I walked the last leg of the journey and arrived home a hair over four days after leaving California.

Apartment hunting in Bowling Green was no more successful the second time around. I settled for a room at the Ross Hotel. The Ross was a rooming house, and I sprang for the deluxe suite. The room was the same as all the others: small, dark, and dismal. However, it had a mini-fridge in addition to a bed, a desk, and a chair. The bathroom was down the hall, and there was a communal cooking area. Each room bore a sign clearly stating, "NO COOKING PERMITTED." It was a rule I would break almost every day I lived at the Ross. I never set foot in the communal kitchen. The Ross had three things going for it: it was cheap, I could walk to campus, and it was so loathsome spending all my time at school would be a better option than my new home.

6. Bowling Green State University

The cure for boredom is curiosity. There is no cure for curiosity.
— Dorothy Parker

New kid and outsider. I was both in the fall of 1971. I joined Pat Muzzall and Mark Eberhard as Master's students. Both did their undergraduate degrees at BGSU, took classes with Rabalais, and knew his program and personal eccentricities. Two PhD students, Jeff Butts and Carol Votava, completed Master's degrees at BGSU with Elden Martin, an avian physiologist, before deciding to pursue parasitology for their doctorates. They were also aware of the strictures Rab imposed on his students. Rab was good to his word: every day, from 8 am to 4 pm and 7 pm until midnight, with Friday or Saturday night off. First semester I was there every night because anywhere was better than the Ross. I had no car, disposable income, or social life. I was petrified of getting even one B and getting the boot. I didn't set foot in a bar for the next two years, and I didn't learn Bowling Green had a downtown until a visit four decades later.

Our group was cult-like. Rab radiated an aura of "us against the world." He wore a short lab coat, cut like a sports jacket, and stalked the halls with his head down and hands, balled into fists, thrust in his pockets with such force they strained the stitching just short of the breaking point. Approach at your peril was the persona he presented to outsiders. Born in central Louisiana to parents of Cajun ancestry, Rab worked diligently to excise any hint of a regional accent and mimicked the neutral tones of his adopted home in the Midwest. He was a rebellious youth, working in the oil fields and riding the rodeo circuit (his specialty was Brahma bulls) before pursuing an academic career. With his students, Rab was flat-out fun. He regaled us with stories of rodeos and bar fights, as well as the scientists whose papers we read for his parasitology course and our research projects.

Science runs on money, and Rab didn't have any. His attempts to secure grant support ended in failure and frustration. "Midnight Shopping" was Rab's solution to the problem. We scrounged for equipment and supplies. We were in the building long after everyone else, at least anyone with any sense, went home to their families. Rab gave us his master key and a list of materials needed to restock his lab or a piece of equipment for an experiment. We raided the laboratories of other faculty to see what we could find. We were cautioned never to take so much it might be missed. Our victims would think they were using supplies a little faster than usual and would purchase more. We were thieves, Fagin's boys from *Oliver Twist*. We were loyal to Rab and each other. We weren't cult-like. We were a cult.

I had no experience teaching, but teaching was how I was supposed to earn my meager $190 monthly stipend. My assignment was the non-majors biology labs under the direction of Reg Noble, an Algologist, who taught the lectures for the course. We met weekly to review the material for the upcoming lab and then set loose on unsuspecting students. I had four sections that met twice a week for an hour at a time. We lectured, did demonstrations, assisted students when necessary, and gave and graded quizzes. Teaching was on top of the courses we took as degree requirements and the additional tutorials Rab held sporadically during the evening before leaving for home. Ninety hours a week, or more, was the norm.

I consider students in non-majors biology as people who have demonstrated by their presence; they would rather be somewhere else. It is a tough sell to would-be novelists and aspiring titans of industry. There were always students in a class of 100 or more who were bright, loved learning, and could have majored in biology, but their interests were elsewhere. It was another hurdle to clear and box to check in the requirements constituting a liberal arts education. Teaching labs was an exercise in deception. I had to convince the students (and myself) into believing I knew what I was doing. Classroom discipline was never an issue at Hiram. The students were primarily middle-class white kids from suburban areas of the northeast and Midwest. We were obedient, if not always attentive. We knew how to play the game. The racial unrest of the late 1960s resulted in many state legislatures mandating programs designed to get African-American students into higher education. Bowling Green was part of the solution.

I had no idea the criteria used to select the young men and women assigned to my classes to attend university. I suspected they performed well in high schools where the bar wasn't terribly high. Many of them were bright and eager but lacked any real preparation for the rigors of college life. I also suspected it was the first time away from home for the majority. There were no (to my knowledge) mentors to assist them in learning how to allocate their time, provide instruction on effective study habits, or adjust socially to what must have been, for many, the equivalent of the far side of the moon.

Each week's class began with a quiz on the material from the previous session. I noticed one young African-American man in the front row, sitting in the middle seat of a lab bench of three, glancing back and forth at his neighbor's papers. I walked over and stood directly in front of him, at which point he focused on his own quiz. He flunked. The following week he took an empty seat at the back of the lab, again in the middle. I placed his quiz at his assigned spot. He moved to the front and shot daggers at me. He took the quiz, failed again, and never returned to class. I felt sorry for him. Neither of us was up to the challenge we were facing, and he paid the price. A more experienced teacher, a more experienced me, would have handled things differently and might have resulted in a positive outcome. Some of those young people undoubtedly succeeded by native intelligence and sheer determination. I also suspect many failed because we failed them.

While Rab's rules might sound draconian, they were what I needed. I required the enforced discipline of a workout in the pool. I needed someone to tell me this is what you have to do if you want this life. I spent 90 hours a week in the lab because I wanted to succeed, or maybe because I was petrified by the thought of failure. Over time my desire to become a scientist increased, and my curiosity to understand how the world worked grew. I earned all A's in my courses, and I started thinking about continuing my education and pursuing a PhD. I began the semester as an outsider. By Christmas, I was part of the group and found a direction and a home.

Life at the Ross Hotel was unbearable. It didn't take long to realize it was primarily for transients; a halfway house for people getting out of the pen was how I described it to family and friends. I cooked all my meals in a small coffee pot in my room. The device served as a double boiler. I bought canned spaghetti and boiling bag meals, removed the lid, stripped the paper off the can, dropped it, or a plastic pouch, into the pot with a small amount of boiling water. In a few minutes, dinner was served. I prepared my evening meals for three months living in fear of getting caught and being banished for breaking the rules. I had to find somewhere else to live.

I asked my parents for their 1966 Dodge Dart as a graduation present from Hiram, and they firmly but politely said, "No." Six months later, they had a change of heart that allowed me to move to an efficiency apartment in an old motel on the outskirts of town. It was a five-mile drive each way, but I had my own bathroom, refrigerator, a 2-burner hot plate, and no restrictions on meal preparation. All those amenities for only $70 a month! Skid row to genteel poverty.

And what about Ms. Fenton? I invited her to visit me at BGSU during the fall. She agreed. I rented a motel room as overnight guests were not welcome at the Ross. Not that I would have wanted her to see the place anyway. We had dinner at a restaurant: a welcome break from coffeepot cuisine. The encounter wasn't what I had hoped. Perfunctory conversation followed, and while we parted amicably, the fate of our relationship was in doubt.

Second semester was a carbon copy of the first: long hours at school teaching non-majors labs and another round of required courses. Two events that semester had a profound influence on my future. First, I learned about summer classes offered at the Gulf Coast Marine Laboratory in Ocean Springs, Mississippi. I read the brochure and was excited to learn they offered a parasitology class taught by Dr. Robin Overstreet. I was dismayed to discover he only offered the course in odd-numbered years, and it was 1972. The only other class of interest was Marine Invertebrate Zoology, taught by Dr. J.J. Friauf. Bowling Green offered competitive scholarships to cover the $500 cost of tuition, room and board. I planned to apply; however, I had little savings, and I vowed I would not ask my parents to support my graduate education. My father sent an occasional check for $25, which I gladly cashed, but it was his choice. I never asked. I completed the application for a summer stipend, but there was no guarantee I would get it.

I was determined to be in Mississippi for the summer, come hell or high water. The awards wouldn't be announced until late in the semester, so I devised a plan to fund the trip myself. I decided to save $70 each month from my $190 paycheck to cover the cost. My monthly budget was $70 for rent, $70 for summer school, $25 for gas, and $25 for food and fun. There was precious little of either. I had to cut down somewhere, and the only place was food. For the next four months, I subsisted on half a can of Dinty Moore stew on a cup of white rice for dinner and coffee — lots of coffee. Breakfast and lunch were luxuries I couldn't afford. Rab was a caffeine addict and provided it in the lab for the students. The only catch was everyone followed his lead and drank it black — no cream or sugar. It kept me awake but provided no calories.

The second event was a Spring Break trip to Cameron National Wildlife Refuge in Cameron Parish, Louisiana. We weren't on a beer-soaked, sun-drenched bacchanalia. We were going to hunt parasites on the southwestern edge of the state. Over seven days, we collected all manner of fish, birds, snakes, frogs, and turtles to go under the knife and yield their worms: 225 animals in all. The only animals off-limits were alligators as they were still endangered and not the swimming pool nuisance and threat to small dogs they are today. However, the refuge staff provided an eight-foot gator, mortally wounded in a territorial battle with an even bigger male, for examination.

The area had been hit by hurricane Camille a few years earlier. Evidence of Camille's wrath was visible in the form of hunks of plywood and sheet metal scattered in the areas where we collected. Flipping the remnants of homes, garages, and sheds was a game of Russian roulette. Any chunk of debris might conceal a water moccasin, an ill-tempered and highly venomous snake. We had snake tongs, three-foot-long poles with a handle and trigger on one end, and a pair of pincers on the other. The objective was to grab the snake behind the head and plop it into a pillowcase without being bitten. It isn't as dangerous as it sounds as long as you maintained a healthy degree of respect for your quarry.

One day, four of us were out collecting. I got out of the vehicle with one of the other guys while our two companions left to explore a different site. We were poking around looking for whatever we could find when I stumbled on a moccasin. I used my snake tongs and snagged a highly pissed-off animal behind the head. Once secured, I realized we didn't have a pillowcase. They were all in the vehicle, and we didn't know when our friends might return. We had two choices, let the snake go or take drastic action. I didn't want to lose the specimen, so my partner took out his knife and cut off the snake's head. As if on cue, our friends returned. I dropped the headless snake (and the severed head) into a pillowcase and returned to our makeshift laboratory. I arranged my dissecting equipment on the lab bench, preparing to perform a necropsy (the dissection of an animal of a species other than its own). I dumped the decapitated snake on the lab table and turned to grab a bottle of saline.

I felt a hard, wet kiss on my arm. I jumped about three feet and turned to see the snake had coiled, struck blindly (What other option did it have?), and was coiled again. A red circle of drying blood marked where the headless animal hit my arm. After a few minutes, my adrenalin levels returned close to normal, and I continued with the dissection, although still a bit unnerved.

The trip offered relief from my self-imposed one-meal-a-day regime. Rab's parents came to visit and stayed for a couple of days. Rab's mother was a fabulous cook and not put off by the somewhat unusual ingredients we provided. She cooked, and we ate many of our specimens — after they gave up their parasites, of course! Ducks, rabbits, and even the alligator went into the pot, and everything she prepared was heavenly. An ocean of gravy accompanied the meat, red beans and rice, and even a few vegetables thrown in for good measure. Returning to Bowling Green and Dinty Moore wasn't going to be easy. We packed up our gear, cleaned our accommodations thoroughly, and headed for a day trip to New Orleans before enduring the 1,000+ mile drive back to northwestern Ohio.

The last few weeks of the semester were uneventful academically. My grades met Rab's exacting standards — nothing below an A. Since returning from Louisiana, I had been processing and identifying the specimens I collected. It was fun. I liked the historical nature of taxonomic work, the challenge of wading through the literature, comparing what you found against species described in the past, and arriving at an answer. Rab knew many of the scientists whose work I was reading and was happy to share gossip about them. Learning something of my forebears' personality quirks added an additional layer of interest to what most people would view as a dry, technical task.

I progressed far enough to begin thinking about a thesis topic. Rab trained as a taxonomist specializing in trematodes (flatworms or flukes) of snakes. Taxonomy is the branch of biology encompassing the identification of organisms, description of new taxa, and classification. Earlier students completed their degrees doing surveys of the parasites of various birds in the area. These studies entailed collecting and dissecting specimens of one host species and identifying the parasites lurking in different organs. The work is historical, descriptive, and the foundation for understanding biodiversity, biogeography, and evolution. Rab and I met and decided a survey of the parasites of local turtles would be a good thesis topic and guaranteed to yield results.

I went to Columbus in April to celebrate Kathy's birthday. I knew almost as soon as I arrived it wasn't going to end well. We had dinner at a Polynesian restaurant, and I told her about my classes, proposed thesis work, and what I hoped for in a future career. She listened politely and told me she didn't think our relationship was going anywhere, and said, "goodbye." I didn't know what to do. I couldn't stay with Kathy. I didn't have the money to pay for a hotel, and I didn't want to return to Bowling Green. I went home.

I arrived in Youngstown in the wee hours of the morning. Since I didn't have a key, I parked in the driveway and fell asleep. I woke with the sun and met my father at the front door on his way to get the newspaper. He looked at me and stared. He didn't say "Hi!" or "What a nice surprise." or even "What are you doing here?" My father scanned me from head to toe, shook his head, and said, "What the hell did you do to yourself?" Despite Mrs. Rabalais' Cajun feasts, I lost about 30 pounds on the Dinty Moore and rice diet. My weight plunged from the mid-180s to the mid-150s.

My parents patiently listened as I explained my weight loss and Kathy dumping me. On the bright side, I also shared my academic success, proposed research project, and plans to spend at least part of the summer in Mississippi studying marine invertebrates. They couldn't do anything about my relationship, but my father was furious I hadn't asked him for the money for summer school. I explained my decision to pay my own way. My father would have paid for the course and was kind of pissed I hadn't asked. If nothing else, I think I earned his respect for my determination to be independent.

I returned to school and received two blasts of good news. The department awarded me a grant to pay for my trip south. My self-imposed deprivation had been entirely unnecessary, and I was back to three meals a day. The second positive note came in the form of the offer of improved housing. A fellow student offered to let me share his mobile home the following year. His parents purchased it for him as an alternative to a dorm room, an apartment with less-than-serious roommates, or, God forbid — the Ross Hotel. It boasted two bedrooms, a kitchen, living room, and a full bath for less than I was paying for the converted motel room. I finished the semester, cleared out of the motel, and packed for summer school.

A few weeks later, I was ensconced in the Deep South for the second time in less than a year. The Gulf Coast Marine Laboratory consisted of a cluster of buildings of various ages, from new to vintage — aka, ramshackle. Our dorm was new and replaced one flattened by Hurricane Camille. It was a big step up from the digs I called home for the past year. Our class of 20 was an eclectic mix of graduate and undergraduate students from across the country: Oklahoma, Massachusetts, and Ohio, among others. We settled in, made initial introductions, and prepared for our first class on Monday morning.

I entered our assigned classroom a few minutes before nine to find a frail old man standing at the blackboard. His left hand tightly gripped his shaking right as he slowly wrote J. J. Friauf in large letters and a font possibly designed by a drunken earthworm. My heart sank as I realized this guy, who at first blush should be in a nursing home, was our instructor. Dr. Friauf started class with a brief introduction. He was from Vanderbilt University and had taught the course at GCML on several occasions. Lecture was from 9 to noon MWF and labs from 1 to 4 pm TT. We would collect specimens from various locations near the lab and use the station's research vessel to collect on Ship and Horn Islands located about 15 miles offshore. There would be a lecture exam every Saturday morning and a comprehensive lab final the last week of class. Finally, he informed us he suffered from Parkinson's disease. Any questions? Despite his infirmity and frail appearance, Dr. Friauf of Vanderbilt was in charge.

During six weeks, I took 225 pages of single-spaced lecture notes. We collected, identified, and classified over 300 specimens over the same span. Lecture exams consisted of long paragraphs in which we had to fill-in-the-blanks to produce a complete and coherent narrative, all manner of other short-answer questions, with an essay question or two thrown in for good measure. Field trips were equally demanding and utterly fascinating. The area around the lab was rich in invertebrates. We brought them back to the lab for observation and identification. We spent afternoons on MWF in the lab, identifying and reviewing specimens. Evenings were for studying for the end-of-the-week exam. It was exhausting, but at the same time, exhilarating. J.J. probably forgot more about invertebrates than I would ever know.

The mental and physical workload stimulated a healthy appetite. Food prepared by the cooks in the dining hall was abundant, of excellent quality, and thoroughly Southern: eggs, ham, grits, fried chicken, potatoes, biscuits, and always, always — lots of gravy. The rules were simple. You could go through the line once, but you could take as much as you could pile on your tray in one pass. Despite the physical and mental exertion, in six weeks, I gained back 25 of the 30 pounds I lost on the Dinty Moore diet.

During the last week of class, we had access to small boats to collect on our own as a review for the lab exam. Each pair of students could sign out a dinghy with a nine-horsepower motor and spend the day exploring the habitats around the lab. My study partner, I will call him Bill because I can't recall his name, and I thought we would take advantage of the opportunity to study and relax a little before the final push. We would get a couple of six-packs of beer, cruise out of sight of the lab, and enjoy the warmth of summer in the south. The day before our planned "Huck Finn" adventure, Dr. Friauf stopped us after class and said he thought he would join us. I don't know if he caught wind of our scheme, but it didn't matter. We sputtered a bit and, with complete insincerity, said we would be delighted to have him along.

We spent the morning and early afternoon cruising the inland waterway looking for likely collecting spots. Periodically we would beach the boat, collect anything interesting, review our identifications, and move on. Shortly after lunch, the inevitable afternoon storm clouds appeared, and we thought we should head back to the lab before the wind really kicked up. Bill was sitting in the front. I was in the back piloting the boat, and J.J. was in the middle seat facing me. We were in a hurry, so I had the throttle wide open. I noticed a slight movement from Dr. Friauf and realized he was going to do something so incredibly stupid it belied belief. And he did. J.J.'s left hand was on the handle of an empty specimen bucket. He picked it up and dropped it over the side. You don't have to be a physics major to guess what happened next.

Newton's Third Law took over. The sudden drag of the now full bucket lifted J.J. straight out of his seat and headed overboard. I held onto the throttle, stood up, put my hand on J.J.'s shoulder, and shoved him down to keep him from leaving the boat. The bucket, ripped from J.J.'s hand, was floating a few dozen yards behind us, but he was safely in his seat. I slowed the motor and circled back to retrieve the lab's property; all the while, J.J. muttered over and over, "I can't believe I did that. I can't believe I did that."

As we approached the end of the class, my grades on the lecture exams hovered in the A− range. I needed to do well on the lecture final and lab exam to get an A in the course. Bill and I were inseparable for the last few days studying 10–12 hours at a time. I wouldn't learn my scores until I returned to Ohio. I aced them both and earned an A.

It was late July when I got back to Bowling Green. I moved into the trailer, which occupied the last spot in the final row of the trailer park, which meant we had no neighbors on two sides. Our front and only door framed blue sky and unending farm fields. It was good to be back, and I needed to get to work on my thesis project if I wanted to finish my degree by the following June, my self-imposed deadline. But there was something I was going to do despite the rules, despite the deadlines.

The U.S. Olympic Swimming Trials were being held at Portage Park, Chicago, in early August. The meet would determine the composition of our swimming team for the upcoming games in Munich. I was going. I told Rab I was taking a five-day break. I was happy to abide by the rules, but I had spent six weeks of almost non-stop studying, and I needed time off. I think he was a bit taken aback because I wasn't asking. His response? "Fine."

On August 2nd, I headed for Chicago. I still loved swimming, and I wanted to see Mark Spitz, the most dominant swimmer on the planet, compete. More personally, a member of my old YMCA team was also vying for a spot on the team. Ross Wales won a bronze medal in the 100-meter butterfly at the Mexico games in 1968, and I wanted to see if he could qualify for Munich. Ross finished a close second in his heat and third overall in the morning preliminaries. In the finals, Ross finished fifth, and his swimming career was over. Spitz won four events, both flys and the 100 and 200 free, setting world records in three of them. He would win seven golds in the terrorism-marred Munich games, a record he held until Michael Phelps won eight golds in Beijing.

Rab arranged for me to collect turtles at the Ottawa National Wildlife Refuge, a 45-minute drive from campus. We met with the staff and scouted locations to set traps — hoop nets. I was going to focus on two species: one common and one rare. Painted turtles (*Chrysemys picta* subspp.) were probably the most common turtles in the United States. The Midland painted turtle, *C. p. marginata,* was the Midwest subspecies and easily identified by variations of scale patterns on the top part of the shell. Blanding's turtle (*Emydoidea blandingii*) was rare, restricted to the upper Midwest, and easily identified by its bright yellow throat.

The easiest way to catch turtles is the hoop net. The design is simple. Take two or three metal rings and encircle them with netting forming a tube. The front ring has an additional cone of netting tapering to a small opening ending about a third of the way through the body of the main cylinder. The posterior ring has a similar cone continuing outside ending in a drawstring. The net was anchored with stakes, so the tube rested on the bottom of the lake or pond and supplied with bait appropriate to the animal you are trying to catch. A turtle attracted to the aroma of a punctured can of tuna dangling from a string on the middle ring swims into the front funnel and proceeds to the small opening at the funnel's end. Once it pushes through, the small opening collapses, and the turtle cannot escape the body of the trap.

My routine involved a daily commute from BGSU to the refuge, check traps, rebait them if necessary, and return. Back in the lab, I dissected my catch, collected, and preserved the worms for staining and subsequent identification. In between, I read everything I could find about the parasites of North American turtles. I scoured our library and submitted dozens of interlibrary loan requests for articles in journals our library didn't carry.

Taxonomic papers (or scientific publications in general) are a cure for insomnia for the lay reader. Measurements of various anatomical structures make up the bulk of papers in the early literature, followed by a brief comparison of the species under study with others previously described and why the current species differed from the others. I found these works fascinating! Why? I have no idea, but I did. Most people don't give parasites a thought when they see an animal in nature, but most harbor one or more species of worm. Most of these infect the vertebrate host through the food chain. Many turtles are omnivores (eating both plants and animals) and typically harbor several different species of trematodes and nematodes in the intestinal tract. A few (notably blood flukes and monogenetic trematodes) infect the host directly via a free-swimming larval stage. They locate in extra-intestinal sites: the bloodstream and oral cavity or urinary bladder, respectively. You may see a squirrel, a robin, or a turtle as a cute little animal. And what do I see? Yep, a bag of worms.

As the fall semester commenced, I made slides, measured worms, and compared my finds to those reported in my growing reprint collection. Reg Noble, my boss for the non-majors biology course, thought I did an excellent job teaching labs in my first year and offered me the position of Lab Coordinator. I was responsible for ensuring supplies needed for each lab were available in sufficient quantity for all of the sections taught during the week. In return, my teaching load dropped from four to two labs per week, and I received a $20 raise.

The tenor of our lab deteriorated during my absence. Rab moved out of his house. He and his wife were getting divorced: an unanticipated event with unexpected results. Shouting matches with his soon-to-be ex-wife consumed Rab's evenings. Everyone was careful to exit the lab adjoining Rab's office if they heard the phone ring or Rab lifting the handset from its cradle. If you didn't move quickly, you were stuck in the lab until the diatribe ended, and those exchanges could last an hour or more. Nobody wanted to have to walk through his office while hostilities were in progress, and there was no other way out. As the semester wore on, tensions escalated, and I began to question my future. I planned to apply to BGSU to pursue a PhD with Rab, but now I was having second thoughts.

I joined the American Society of Parasitologists shortly after arriving at Bowling Green. A large part of any scientific society is an annual meeting where like-minded folks gather at some exotic (by my standards) location, drink beer, and share the results of their recent research. The next meeting was in Miami, Florida, in early November. Pat Muzzall, Dave Ashley, Mark Eberhard, and I decided this was too good an opportunity to let slide. We paid the registration fee, and in early November, piled into a car for the 1,300-mile trek south.

My primary objective was to meet Horace Wesley Stunkard, one of the founders of the society who had written a series of papers on turtle parasites in the 1920s. Stunkard was over 80 at the time, and I wanted to meet him before he died. Little did I know he would live to be 100, and I would converse with him on several occasions in the ensuing years.

The meeting was a revelation. We met the people whose names graced the papers we read for class and research. It was akin to touring Hollywood with a map to the stars' homes and being invited in for coffee or a beer. We attended lectures, listened to scholarly presentations, and heated discussions in the meeting rooms and hallways. I was much too intimidated to join in and was content to be a fly on the wall.

The meeting was coming to a close, and I hadn't met Dr. Stunkard. Barroom chats with older graduate students painted a somewhat frightening picture of the octogenarian. In his younger years, Horace's reputation included ruthless questioning of graduate students presenting their PhD work. The stories I heard suggested he brought more than one of them to tears. I screwed up my courage and, with extreme trepidation, introduced myself as he sat, alone, in the hotel lobby. Dr. Stunkard was tall, thin, impeccably dressed, in excellent physical condition, and mentally alert for someone his age. He patiently listened as I babbled about reading his work, my fascination with turtle parasites, and some other nonsense I don't recall. We talked for a few minutes, and I said, "Goodbye," feeling good about the encounter. Perhaps the assessment of one student I spoke to was accurate; Stunkard's caustic demeanor had softened over the years because "He wanted to die with at least a few friends."

Attending the ASP meeting and the constant hostility surrounding the breakup of Rab's marriage strengthened my resolve to find somewhere else to do my PhD. Of course, I would need a letter of recommendation from Rab for my applications. I shared my plan, if not all the reasons for leaving, and he graciously complied. I identified two people I thought I might like to work for: Martin Ulmer, an excellent taxonomist at Iowa State University, and John Holmes, a parasite ecologist at the University of Alberta (Canada) and most recent recipient of the Henry Baldwin Ward medal as our society's Outstanding Young Parasitologist. I completed the applications and put them in the mail a few days before Christmas. Why pursue a PhD? At some level, I must have had some thought of a career in academics, but it wasn't the prime motivator. I wanted to see if I could do it. Did I have the intellectual wherewithal to compete in the most rarified air of academic pursuit? Only later did I learn completing a PhD is more a matter of persistence than intelligence.

Second semester was devoted to writing my thesis. I was chafing under the strictures of the 8 to 4, 7 to midnight rule. I didn't need to be in the lab 13 hours a day, 6 ½ days a week. Rab wasn't living up to his definition of a major professor as his family situation occupied most of his time. More times than I care to recall, I sat reading the same paragraph of some random textbook over and over and over, waiting for midnight. Jeff Butts took things to the point of anarchy. Jeff watched the parking lot from a window on the third floor every evening at ten sharp as Rab's headlights disappeared into the darkness. Then Jeff bolted out the door to spend time with his wife. I wasn't alone in my rebellious thoughts.

In mid-March, I received notification of my acceptance into the PhD program at the University of Alberta with a full scholarship. John Holmes would be my advisor if I accepted their offer. I knew little about Canada other than from family vacations when I was a kid. Alberta and Edmonton, home of U of A, were black holes. I hadn't heard from Iowa State, so I called Dr. Ulmer to see if he could shed some light on my status. Dr. Ulmer carefully listened as I related the situation. He hesitated for a moment and said he recalled my name but not my application. He asked if he could call back in an hour. I gave him the number and waited. An hour later, the phone rang, and Dr. Ulmer explained his confusion. My application had been received but had fallen between two filing cabinets. The graduate committee met earlier in the week and awarded fellowships for the following year. I was not among them. Martin's solution was for me to start in the fall, pay my own way for the first semester, and then he could probably get me money for the spring. I thanked him for his candor, hung up the phone, and started looking for information on Edmonton. I was going to Canada!

My thesis was nearly complete; all I had to do was prepare for my oral examination. One of the other students in the department received an acceptance letter to medical school, and he and his wife were throwing a party. It wasn't a BYOB either. They would supply the food and drinks. I hadn't attended a party for nearly two years, and I was excited as hell. I arrived a little after 7 pm, grabbed a beer, some food, and began circulating. I hadn't consumed half the bottle when the host's wife tapped me on the shoulder and told me I had a phone call from Rab. Holy shit! I couldn't imagine what he wanted, but I wasn't going back to the lab — no ifs, ands, or buts. I picked up the phone. Before I could be indignant, Rab said someone wanted to talk to me. It was Kathy. I told her I would be right over. I put down the beer, expressed my regrets to the host and hostess, and headed back to campus.

My only contact with Kathy for the past year was a postcard from Munich sporting a photo of a partially constructed pool, the site of the swimming competition for the terrorist-marred 1972 Olympics. Otherwise — nothing. The drive took five minutes, and I bounded up the three flights of stairs of the Biological Sciences building two steps at a time. Kathy was at the midpoint of the long hallway. As I approached, I looked her in the eye and said, "What the hell do you want?" Not exactly the greeting to rekindle old passions. I hurt her, and it showed. That is not who I am. I apologized; she recovered and told me she had come to say goodbye. Kathy shared that teaching junior high physical education at an inner-city school had taken its toll, and she planned to move to San Francisco with a friend. Our conversation became more congenial, and I suggested we go back to the trailer, where we would have some privacy. I had no ulterior motive; I wanted to get out of the building and away from Rabalais.

We talked for hours and shared our lives during the past 12 months. The mutual attraction was building with each passing hour. Around 3 am, Kathy said she should leave before we did something we might both regret.

I assured her I would behave like a gentleman; however, she insisted, but not before we arranged to meet in Columbus after my thesis defense, which was a month away. She left, and I was in love. I thought she was too.

What happened? I grew up. My reluctance to deepen our relationship in the past was immaturity, the proverbial fear of commitment. In the past two years, I changed. I accepted Rab's draconian rules and thrived. I committed to parasitology and, I hoped, a career in science. I did hard things and succeeded. I loved this woman. I knew it wouldn't be easy. But I wasn't afraid to try something new, something challenging.

April disappeared in a heartbeat. I finished my thesis, and it passed muster with Rab. The only barrier to completing my degree was the oral exam. My committee included Rab, Bob Romans, a botanist who had become a friend and confidant, and Richard Baxter. I had taken Immunology and Biostatistics with him, and we had a good but not personal relationship. I paced the hall waiting for the 1 pm start time. I was a wreck. Rab assured me I knew the material in my thesis cold — better than anyone else in the room. His advice was simple. Listen carefully, pause to consider what I wanted to say, and, most importantly, answer the question posed. Two hours later, the three men looked at each other and congratulated me on an excellent thesis and clear, concise answers to their questions. I flubbed a few but not enough to even remotely give them pause.

In late May, I headed for Columbus. A few of the old fears started to resurface, but I damped them down. I wasn't going to screw this up again. Kathy was house-sitting for the parents of one of her roommates in a posh suburb. No shabby student digs for this rendezvous. We spent most of the weekend in bed. We didn't have sex; we made love. I proposed after one round of intimacy, and she accepted enthusiastically. The only problem was when. She and her travel buddy, a friend from Hiram, had already planned a two-month trip through the western United States. She could not conceive of pulling off a wedding before early fall — a serious problem. I was supposed to be in Edmonton in late August.

Before I left to pack and depart Bowling Green for my summer job lifeguarding and coaching swimming at a local country club, I promised to do everything I could to push back the start date of my PhD until January.

I wrote to Dr. Holmes and explained the situation. He was concerned but incredibly gracious and told me he would work things out with the university. He could have told me to get my ass up there or lose my fellowship. John is one of the kindest people I know, and it wasn't the last time his compassion helped shape my life.

I returned to Youngstown and immediately started looking for an apartment for the summer. After living on my own for six years, the thought of adhering to any rules my parents might want me to follow, however lenient, wasn't going to work. Kathy left for her western excursion, and I settled in for a summer of country club life. I was recently engaged, my fiancé was hundreds of miles away, and I hadn't been with anyone but Kathy (for one all too short weekend) in two years. Temptation, in the form of members' daughters, wait staff, and other lifeguards, was all around me. It would have been easy to stray, but I am glad to say I didn't then and haven't since. As Dr. Seuss' character Horton repeats in "Horton Hatches an Egg," "I meant what I said, and I said what I meant, an elephant faithful 100 percent."

My departure from BGSU was amicable, although I had lost some respect for Rab during that second year. He gave me the opportunity to learn discipline in pursuit of a goal and a good grounding in parasitology. The cult-like atmosphere dissipated, and our tiny band of worm hunters began to drift apart as we pursued our own futures. Like Camelot, we experienced a brief moment with like-minded souls, but it wasn't meant to last. Pat Muzzall finished and went to the University of New Hampshire for his doctorate. Mark Eberhard did his PhD at Tulane, and I headed for Alberta.

Carol Votava and Jeff Butts both finished their PhDs with Rab. All, save Carol, landed positions in parasitology and had successful careers. Carol returned to Bowling Green and became Head of Nursing at a local hospital. She and Rab married shortly thereafter. Was there something between them earlier? If there was, I missed it.

Rab never had another group quite like ours. He never got the 'big' grant he wanted so desperately. When Rab failed to earn promotion to Full Professor, he pretty much quit. He started a business raising and training Doberman Pinschers, only coming to campus to teach his classes and play cards. He and Carol retired and moved to Tennessee, where Rab had been buying land for years. I saw him once or twice at parasitology meetings; the conversations were polite and perfunctory. Rab died at the age of 73 from complications due to Alzheimer's. His only enduring contribution to parasitology was us.

7. The University of Alberta

*The seeker embarks on a journey to find what he wants and discovers,
along the way, what he needs.*
— Wally Lamb, The Hour I First Believed

Kathy and I were married on a glorious day in early October 1973, with a reception held in the backyard of family friends. The sky was clear and blue, with temperatures in the low 70s. Kathy's mother tracked the weather on October 6th for the next decade, and it was mostly miserable. An omen? Perhaps, if you believe in those things. I don't. We shuttled between both sets of parents and Kathy's sister's and brother-in-law's home in Jamestown, New York, for the next six weeks before our departure for Edmonton in late November.

We had one major hurdle to clear — Canadian Immigration. There was voluminous paperwork to complete and medical exams to demonstrate we were not bringing any infectious diseases with us. The final step was an in-person interview at the Canadian Consulate in Detroit. We arrived in the city at approximately noon for our 1 pm appointment, parked, and had lunch. The interview room was small with four chairs, two for us and two for the consular officials. They had one question. How much money were we bringing with us? I didn't have two nickels to my name, but Kathy was a saver. Her bankbook showed a balance of over $6,000, enough to suggest we would not go on the dole if things got rough. The entire process took less than 15 minutes.

We were granted Landed Immigrant Status and were back home for dinner. Being Landed Immigrants was huge. It meant Kathy and I had all the rights of citizens, except we couldn't vote or hold public office. Kathy could work in the private or public sector and compete with Canadians for positions on an equal footing. Most students received student visas, and neither they nor their spouse could earn money beyond their university stipend. We made out like bandits during our 4 ½ years north of the border.

Our first impression of Edmonton was thoroughly depressing: industrial, barren, and ugly, as exemplified by a 30-foot tall cowboy advertising a tire dealership we passed as we headed toward Michener Park, the married student housing complex south of the University. Our apartment in a high rise wouldn't be vacant for ten days. We had to camp out in an empty unit in one of the low rises in the interim. We signed some forms, got the keys, and moved our meager belongings into our short-term squat. The apartment was bare save for a few kitchen appliances and a single twin bed scrounged from a storeroom known only to the staff. I spent the next ten nights sleeping with my back against the wall before purchasing a queen bed set and moving into 'our' apartment in Vanier House.

Dinner that night was downtown at the Old Spaghetti Factory. As we approached the Athabasca River and crossed the High Level Bridge, our hearts stopped, our eyes widened, and our jaws dropped. We were on a bluff overlooking downtown, and the view was Oz-like, only painted with a more vibrant palette. The Capital Dome glowed in the distance, and the city, in the semi-darkness of early winter, was breathtaking. We soon discovered that Edmonton, then half a million strong and miles from anywhere, was a dynamic, energetic community and a prime destination for performers from around the world. The populace was diverse, affluent, and starved for entertainment. Almost any performer could command an audience of several thousand eager fans. Edmonton boasted two professional theater companies and a world-class symphony.

The following day we went to the 'Uni' to meet Dr. Holmes, enroll and get oriented to the Zoology Department. Things went smoothly, and we finished by early afternoon. Since we had no food at our temporary digs, a trip to the grocery store was on our to-do list. As we proceeded up and down the aisles, Kathy reminded me it was Thanksgiving (American version), and we were having turkey for dinner — no ifs, ands, or buts. The problem was the only birds available were frozen. No matter, a ten-pounder went into the cart. Since we only packed the bare necessities in the car, Kathy couldn't bake a pumpkin pie, so pumpkin ice cream was a suitable substitution. We returned to our place, unloaded the groceries, and filled the bathtub with hot water to thaw the bird. We enjoyed our first Thanksgiving meal as a married couple at 10 pm. No matter. We were young, in love, and starting the adventure of our life together.

The next few weeks were hectic. We purchased a bed and sofa and moved into our apartment in the high rise. The moving van appeared, and we spent the next few evenings decorating our new home. The apartment was beautiful and, like so many beautiful things, utterly impractical. The building was hexagonal, with six apartments per floor. The entryway was short with a shotgun kitchen on the left, a coat closet, and stairs leading to the second level on the right. The remainder of the 1st floor was a vast living area terminating in acute angles on both sides, which resulted in a great deal of unusable space. The center of the outside wall was interrupted by sliding glass doors that should have led to a balcony. During construction, the building was over budget, so they scrapped the balconies and installed a protective railing to prevent the occupants from falling out when the doors were open. The second floor consisted of a bedroom, bathroom, and an open space that was pretty much useless. We loved it anyway. It was our first home.

I was eager to get started, but classes didn't begin until after the first of the year. Dr. Holmes had a mix of graduate students: those starting (Al Bush, Randy Vaughn, and me) and others well into their programs (Ray Leong, Chandra Sankurathri). Jay Hair was a mysterious figure. He was doing a PhD with John but had a teaching job in the States and only came north in the summer to do research. Ray was from Malaysia and studied parasites of various fish species in Cold Lake on the Alberta-Saskatchewan border. Ray was planning a collecting trip the week before Christmas and needed some additional bodies for grunt work. I couldn't wait. Norm Williams, a student of Jerry Mahrt's, the department's Protozoologist, completed the crew.

Ray requisitioned a department truck, loaded his gear, and we headed east. The university supplied a trailer as living quarters for students doing fieldwork at sites too far to commute, our home for the next few days. Typically, Ray would haul all of his equipment onto the lake in the truck;

however, it was early in the season, and the ice wasn't thick enough to support a two-ton vehicle. We had to do it the old-fashioned way, on a large sled similar to those used in the Iditarod. The weather was clear and cold, −25°C. The sun was blinding, and the sky a brilliant shade of blue I had never seen. Norm and I were the sled dogs and hauled the gear between a quarter and half a mile from the shore. Ray grabbed an ax and chopped a square opening in the ice. He picked up a long, thin board with a stout metal hook in the center. Ray tied a long rope to the hook so the board wasn't lost when submerged under the ice. The board was painted bright yellow and clearly visible through a foot or so of frozen water. When Ray pulled the rope, the tooth came up and engaged the ice, pushing the board down. When Ray released the line, the board shot forward about ten feet; he repeated the process until it traveled about 100 feet from the starting point. Ray got the ax, chopped another hole, pulled the board out, and detached the rope. Ray tied his end of the rope to a fine filament gill net, and Norm and I began pulling it under the ice closer and closer to our opening.

We secured the net to the ice and started packing the sled for the return trip. I was busy loading equipment when I heard a cry for help. Ray backed into one of the openings and was floundering waist-deep in the water clutching the edge of the ice to keep from sinking to the bottom. Norm and I pulled him out, pushed the gear off the sled, and strapped Ray in. We covered the distance to the trailer faster than seemed possible. By the time we arrived, my lungs were burning from inhaling the frigid air. We got Ray inside, stripped him down, cranked up the furnace, and wrapped him in blankets. He was fine, embarrassed, but fine, and I had my introduction to the dangers of working in the north.

Our night in the trailer was an adventure. The folks who built the box insulated the ceiling and walls but not the floor. A distinct thermocline developed at waist level. When standing upright, your breath suddenly condensed, forming a cloud as it sank past belt level. The top bunk was almost tropical, requiring only a light blanket, while the occupant of the bottom bunk might as well have slept outside and needed a down sleeping bag. To call our accommodations Spartan would be generous. It had a sink, a stove, fridge, and running water, but not much more. The 'facilities' were outside. You didn't go unless absolutely necessary. Before my first excursion to the outhouse, Norm warned me if I had to sit, to sit on my hands. Otherwise, my butt cheeks might freeze to the toilet seat. I wasn't sure if I was being had or not, but I wasn't taking any chances. I sat on my hands.

The following day was a carbon copy of the day before: cold and clear. The hike to the nets was unhurried as we scanned the lake and its surroundings. Breathtaking didn't do it justice. Ray took the ax and broke open the thin veneer of ice covering each opening. We untethered the ropes securing the nets at each end. I started pulling while Norm maintained the tension on his end. As the net emerged from the water onto the ice, the fish, trapped by their gill covers, struggled in a futile attempt to escape. Ray expertly removed each animal and dropped it onto the ice. At 25 below, the fish flopped two or three times before their gills froze, and they lay still. The catch topped three dozen, mostly whitefish, by the time the last frames of the net left the water. We packed our quarry in coolers and covered them with ice for the trip back to Edmonton, where Ray would necropsy each animal, carefully counting, preserving, and identifying the parasites present.

The University of Alberta functions on the British model of a research institution. Advanced degrees required no set number of course credits, save those mandated by your advisor (generally theirs), and to address deficiencies identified during the qualifying exam. Dr. Holmes taught a year-long course titled "Helminthology." Each week covered a different group of worms. Weekly assignments included 800–1,200 pages of readings, discussion questions, and miscellaneous items, depending on the group. We met for three hours on Friday afternoon to review the material, argue points raised in the readings, and answer questions posed by Dr. Holmes. Reticence was not considered attractive, and the debates could be intense. For a year-end project, each student was required to write a paper describing a new species of parasite, in the form of a manuscript to the *Journal of Parasitology*, a holdover from Dr. Holmes' graduate student days at Rice University with Asa Chandler.

The 9th floor of the Biological Sciences Building was all parasitology all the time. There were three faculty: John Holmes, Bill Samuel, and Jerry Mahrt. Bill was a wildlife parasitologist, and Jerry a Protozoologist. John, an ecologist and theoretician, was the group's guiding force. Each floor had three student labs with four student carrels in each lab. There were also faculty offices with labs and a few singles for students in the final throes of writing. All the student carrels were occupied when I arrived, so John let me set up shop in the lab adjoining his office.

John, while kind and generous to a fault, was, like many academics, an introvert. He didn't go out of his way to talk to people, including his students. Since I was only a few feet away, John talked to me if he needed a break. I wandered over to the student carrels, but I wasn't there for the bull sessions, jokes (practical and otherwise), or when they planned to go to the bar, although nights out with the boys held no allure. I decided early on I wanted to stay married. When I finished for the day, I went home. Despite my isolation, I heard rumblings of discontent with Holmes. Students felt ignored and not welcome in his office. The shit was going to hit the fan; only I didn't know it.

Ski-week, Albertan for spring break, was fast approaching. Our group, students and faculty, was going to Friday Harbor Marine Lab, on San Juan Island, off the Washington coast, for a week of parasite hunting. A reprise of the trip to Louisiana with Rab's group, only our target would be fish. The lab had boats equipped to net fish from different depths to sample the parasites of animals frequenting the surface, mid-water, or deeper environs. We spent our days collecting fish and their parasites.

One evening, unbeknownst to me, the other students invited Holmes to our cabin. I didn't know what he thought might be going on, but I am sure it wasn't close to what transpired. John sat in the center of a semicircle while his students aired their grievances — a Festivus moment if there ever was one outside of *Seinfeld*. John, true to form, listened carefully and agreed to make changes. Despite John's best attempts, several students left at the end of the semester. Randy Vaughn returned to North Carolina and became a veterinarian. Al Bush left but returned a year and a half later to complete his PhD. Bill Igo, a country boy from West Virginia, decided more graduate work wasn't for him. He obtained a position with West Virginia Fish and Wildlife, got married, and had several children. Thinning the herd opened several grad student carrels, and I moved across the hall where I should have been from the beginning.

The winter of 1973–74 was frigid with above-average snowfall and temperatures routinely −20°C or lower. Plows piled snow so high in the middle of four-lane roads they formed nearly impenetrable crash barriers.

Kathy and I didn't discover speed bumps in the parking lot of Michener Park until the underlying ice melted in mid-May!

Kathy found a job at a daycare center, and the arrival of spring increased her desire to spend more time outdoors. And then there was the issue of a dog. I was spending a lot of time in the lab, and Kathy was lonely. One evening in mid-February, we decided to take a walk in a subdivision bordering married student housing. The temperature was well below zero, so we donned down jackets and long johns. The conversation was light, and we were enjoying the cold, clear night sky until Kathy asked, "How long do you think it will take you to finish?" "About four years," I replied. She hesitated for a moment and burst into tears, which started to freeze on her cheeks almost immediately. She envisioned two. Her reasoning was simple; if it took two years for a Master's, it should take two more for a PhD. We were 2,000 miles from family and friends, and two years was, in her mind, doable — four was not. Then came the dog, not permitted in married student housing, the desire to be outside when possible, and the decision to find other accommodations.

I had pressing issues of my own. I needed to identify a thesis project and was meeting resistance from Dr. Holmes. I was attracted to John's lab because of work he did at Friday Harbor. John conducted an elegant study of habitat segregation of two species of blood flukes in rockfish: different species of trematodes living in distinct parts of the circulatory system. One of the species was new, and I was gravitating more and more toward taxonomic work. Unfortunately, John's interests shifted in a different direction: mathematical modeling of parasite communities. Regrettably, I was (and still am) a mathematical ignoramus. Despite Wendell Johnson's best efforts at Hiram, *The Calculus* never took hold. I decided to try again and enrolled in a calculus class for the fall semester.

We had to give two months' notice to vacate our apartment. The new occupants had time to give notice where they were living, and they could move without penalty. In six weeks, we would be homeless. Apartment hunting began in earnest in mid-July. The problem was the vacancy rate in Edmonton was less than 1% due to the oil boom occurring as development of the tar sands got underway. Cars in Edmonton sported bumper stickers proclaiming, "Let the eastern bastards freeze in the dark," as oil became a scarce commodity everywhere but Alberta.

We looked at several places permitting pets, and all of them were either overpriced, dismal, or both. We were running out of time when we thought — why not buy a house? A neighbor, a few floors up, was a licensed realtor. We found a small, 700-square-foot house southeast of campus (10540 63rd Avenue) we could afford. We made an offer of $23,800, which was accepted, and we moved in late August. Both sets of parents opposed the purchase. They were concerned about the mortgage payments, and I would be too busy with school to care for it properly. In hindsight, it was the best decision we ever made.

Getting a dog was one of the main reasons for leaving married student housing. Once we got settled, or at least partially settled, Kathy gently reminded me looking for a pooch was something we needed to do — now. She wanted to head straight to the pound and get one. I suggested a little more research. A friend at BGSU had a Keeshond everyone loved. She was the right size (35 pounds), even-tempered and gorgeous. Kathy had more affinity for mutts. One Sunday, the paper carried an ad for half Keeshond puppies. My only question was, "What is the other half?" The problem with looking at puppies is nine times out of ten people come home with one despite the promise they were only looking.

The other half was Irish setter. The owner of the Keeshond had been waiting for a male to clear quarantine at the border so the two purebreds could produce puppies worth hundreds of dollars each. A horny Irish Setter a few doors down jumped the fence while the female was in heat, and nature took its course.

Keeshonds are compact, with an upturned tail, a short face, and long hair: white at the base, grey in the middle, and tipped in black. Irish setters are, well, Irish setters. The puppies were cute as hell — more Keeshond than setter. Kathy fell in love with a shy female who hung back from her littermates, and we headed home. We wanted a name for our new family member, reflecting our adopted home — Canada. After ten days, the dog was in danger of responding to either Sweetie or Cutie Pie. Kathy called me at the lab one afternoon, and the name issue came to the fore. I pulled out a topographic map of Alberta from one of the lab drawers looking for indigenous landmarks and stumbled across 'Shunda Creek.' Our cute little companion finally had a name. Later research revealed Shunda meant muskeg or swamp in the local First Nation language. I don't think she cared. Shunda was an excellent addition to our family and a loving companion to our children when they came along.

Fall semester was busy with teaching, calculus, and working on thesis proposals for Dr. Holmes. Kathy landed a job as head of the first after-school daycare center in the city. The position paid much better than being a worker bee in the infant's only facility where she was employed. The *Edmonton Journal* took note of this innovative program and featured her in an article when the center opened its doors.

The house was a godsend. It needed a lot of work: repairing windows, painting (inside and out), and many more modest tasks. I even built a study in the basement. Since I had no training in any of these pursuits, I learned by doing. The local hardware store, two blocks away, was my second home. I found the work great therapy from the rigors of my academic pursuits.

I didn't take to calculus (or it didn't take to me) any better than I did with Wendell Johnson, so I dropped the class. My qualifying exams went well, although I was deemed deficient in physiology and had to take a course for remediation. I submitted three different proposals for my PhD research, and Holmes rejected them all. I was frustrated and began to wonder if I hadn't made a colossal mistake. I still wanted a PhD, and I wanted to stay in parasitology. I thought about what I liked and decided transferring to a different university working with a top-flight taxonomist was the way to go. I considered several possibilities and zeroed in on Dr. Gerald Schimdt at the University of Northern Colorado in Greeley.

Dr. Schmidt was the leading taxonomist in our society. He regularly published on new tapeworms, roundworms, and acanthocephalans. I wrote to him, expressed my interest in joining his lab, and inquired if he was taking students. He replied in the affirmative, and we arranged to meet in Greeley to discuss the move. During 'Ski Week,' Kathy and I headed south for a visit to Colorado. Classes were in session at UNC, so Dr. Schmidt sandwiched our meetings between his lectures, labs, committee meetings, and student appointments. Gerry was generous with his time and candid in his assessment of my situation. After two days, his advice to me was simple: stay where I was, if at all possible. He judged me an able student and would welcome me to his lab; however, resources at UNC were limited. In his opinion, the University of Alberta offered more opportunities than were available with him. It was not what I wanted to hear, but I took his advice to heart as we retraced our route back north to Canada and an uncertain future.

I did two things before leaving for Greeley. I told Dr. Holmes I would not be returning in the fall and shared my tale of woe with Bill Samuel. Bill was younger than John by a decade and less intimidating. He was the sympathetic ear for several students on the 9th floor when they needed one. After listening carefully to my story, Bill said he would take me as a student. On the trip back, I decided to put him to the test. It is one thing to make an offer when you are sure the answer will be 'No' and another when it might be 'Yes.'

Bill was in his office when I returned to campus. I asked him straight out, "Were you serious about taking me as a student?" He thought for a moment and replied he was, but I had to clear the move with John. Certainly a reasonable request. I walked to the end of the hall and knocked on John's door. I related my conversations with Dr. Schmidt and his recommendation, our divergent interests, and Bill's offer. John thought for a moment, nodded his head, and said he thought it was a good idea. I had his blessing; however, there was a problem. During our sojourn to the States, the graduate committee met and determined stipends for the fall. John told them I would not be returning. I wasn't getting any money: not devastating but concerning nonetheless. What happened next took me by surprise. John said, "Let me see what I can do. I'll get back to you by the end of the day." An hour later, my stipend was restored, and I had a new PhD advisor. John stayed involved in my career as a member of my PhD committee and a friend for the last 40 years.

I needed a thesis topic, and Dr. Samuel had one burning a hole in his pocket. *Parelaphostrongylus tenuis*, or the meningeal worm, is a nematode parasitizing the connective tissue (or meninges) covering the brain of white-tailed deer. The worm causes no harm to deer, but if it infects moose, elk, or other cervid relatives, there is severe neurological damage manifested by weakness in the hind legs, blindness, an inability to feed, and eventually death. Roy Anderson, of the University of Guelph in Ontario, Canada, was the doyen of the meningeal worm. Dr. Anderson stabilized the worm's taxonomy, elucidated the life cycle, and demonstrated its pathological effects in a series of elegant papers in the 1960s.

He and his students discovered a common and inconspicuous slug, *Deroceras laeve*, transmitted the infection to the mammalian host. He was also instrumental in mapping the range of the parasite, which extended from eastern North America to the woodlands of Saskatchewan (Canada).

In the 1930s, a related parasite (*Parelaphostrongylus odocoilei*) was described from the dorsal musculature of black-tailed deer (*Odocoileus hemionus columbianus*) in northern California. Bill and John examined fecal samples of mule deer (*Odocoileus hemionus hemionus*) from central and western Alberta and found larval nematodes consistent with the genus *Parelaphostrongylus*. However, the larvae could not be used to identify the species present, and molecular identification was decades in the future. My project involved identifying the species present in Alberta, completing the life cycle, and identifying the mollusks transmitting the infection. Bill thought working in Jasper National Park, a 225-mile drive oneway, from Edmonton would be ideal. The resident herd of deer was known to be infected, and they congregated in the forest adjacent to the highway south of Jasper.

I began making the trek to the park in late June. My goal was to collect deer feces and search under rocks and logs for slugs and snails. Fecal samples were returned to the lab, wrapped in cheesecloth, placed in a stoppered funnel containing water, and left overnight. In the morning, I opened the stopper and collected a small volume of water in a centrifuge tube as the larval nematodes exited the mucus coating on the pellets and sank to the bottom of the tube. After a quick spin in the centrifuge, I discarded the bulk of the water and poured the small amount remaining into a Petri dish for examination with a dissecting microscope. A positive sample yielded dozens to hundreds of wriggling worms.

Slugs and snails were treated differently. I minced the body (after removing the shell from snails) with fine scissors and placed the pieces in a centrifuge tube containing digestive enzymes. The tubes containing one slug or snail were incubated in a water bath at body temperature to mimic conditions in the stomach and small intestine. Again, I microscopically examined a small sample of fluid from the bottom of the tube. Worms surviving the process should be infective to deer.

Bill arranged with Alberta Fish and Game to obtain orphaned deer fawns, who lost their mothers in motor vehicle accidents, as experimental hosts. We housed deer at the 'Farm,' a facility on the outskirts of town. Once we had a few fawns, I collected infective larvae from a batch of snails. Technicians placed the larvae in the first bottle of milk the deer received each day. Then I waited. After 40 days, I began a daily examination of each animal's feces. The presence of wiggling worms would indicate an established infection. A little less than two months after receiving the infective larvae, baby nematodes appeared in the poop! Now I had to find their parents.

Finding adult worms was no simple task. We guessed the adults would either be in the central nervous system (the home of *P. tenuis*) or the musculature (*P. odocoilei*), although a new species wasn't out of the question. We choose Mule Deer #3, affectionately known as Harry, for examination. Harry was small and passing a large number of larvae, suggesting the presence of multiple adult worms. One morning in late summer, Bill and I went to the farm. There is no easy way to say this. I killed Harry. Bill wisely forbade my involvement in feeding and rearing the fawns. He knew this day would come, and he didn't want me emotionally attached. Killing Harry wasn't easy but necessary.

There were no worms in the brain, so we ruled out *P. tenuis*. The musculature was another matter. Even though Harry was small, there was still a lot of meat. Each muscle had to be identified and shredded, strand by strand. Mule Deer #3 would become a giant Harry burger. Several students, Ray Leong, Howard Samoil, and Chandra Sankurathri, joined Bill and me in the search. We skinned Harry's carcass and divvied up the muscles. Several hours later, someone (I wish it was me) yelled, "I've got something!" A long, black stripe lay in stark relief on a background of red muscle. Adult worms feed on blood, which turns black when digested. The line was the intestine of my worm. The worms congregated in the psoas muscle, in the small of the back, the exact location for *P. odocoilei* in the original report from California. The first step was complete.

Our house was taking shape. I finished most of the painting and major repair work and became pretty good at hanging wallpaper and other remodeling chores. Kathy applied for a job teaching Physical Education at a local middle school. The principal, Johnny Bright, an American transplant, had been an All-American football player at Drake University in Iowa and played professionally in Canada. She was well-qualified for the position, but being American and blond probably didn't hurt. Kathy's salary, a hair over $20,000 a year, was almost double her daycare center income. Combined with my annual stipend of nearly $6,000 we made good money and saved a minimum of $500 a month. Compared to the other grad students, we were rolling in it. Information we kept to ourselves.

An unexpected note appeared in the grad student mailboxes during the summer of 1975. Someone discovered a PhD in Zoology required proficiency in a foreign language. The department failed to communicate this tidbit to the current crop of enrollees, so they waived it for students already in the program. I decided to do it anyway. Russian scientists made substantial contributions to the literature of *Parelaphstrongylus* spp. and related taxa — all in their native language. I decided to learn Russian.

I scanned the offerings for the fall semester, found Russian 101, and enrolled. The professor appeared at the first class and informed us he was 'not' the instructor for the course but would fill in until 'she' arrived on campus sometime the following week. He introduced the characters and sounds of the Cyrillic alphabet, a few simple words, and the basics of sentence construction while awaiting our 'real' professor. It turned out 'she' was from Czechoslovakia, not Russia, chewed gum while lecturing, and considered herself hip in some eastern European sense. Ms. Hotsie Totsie appeared in class wearing a red tank top and a pair of gold lamé pants so tight they looked painted on. After two classes of unintelligible, gum-popping blather, I knew I wasn't going to learn much of anything. I went back to the university catalog and found the course I should have signed up for in the first place, Scientific Russian, a graduate class designed specifically for people like me. The only problem was the drop/add deadline had passed. Undeterred, I went to the faculty member in charge to plead my case. Fortunately, he was sympathetic and pulled whatever strings necessary to get me out of Ms. Hotsie Totsie's class and into his.

There were four of us: two mathematicians, a physicist, and me. First semester focused on vocabulary, grammar, and other basics. The presentation was clear, and questions answered without any popping gum. By the end of the semester, I had a reasonable grasp of the fundamentals. Second semester, we met individually with the professor one hour a week to review articles pertinent to our research we translated since our last session. The more material I translated, the more adept I became at deciphering the nuances of the language. By the end of the semester, I translated over 200 pages of Russian work on *Parelaphostrongylus* and its relatives. Unfortunately, language retention is not like riding a bicycle; it requires constant attention and use. Within a few years, my hard-won knowledge evaporated. Russian is no more intelligible to me now than Rorschach blots.

I spent hours at the microscope studying the worms retrieved from Harry and a few other fawns we dissected to obtain additional specimens. I became an expert in nematode anatomy. I measured various structures and compared my worms with descriptions in the literature. Drawings documenting the worms' critical features were first made in pencil and then in India ink for publication. I was pleased with my progress, as was Dr. Samuel.

Zoology Department seminars were the highlight of the week. Every Friday at 4 pm, the main lecture hall, with a seating capacity of 250, filled with faculty and students to hear a presentation by an eminent biologist. Some were external examiners, flown in to participate in the thesis defense of a student finishing their degree. Others were selected because they were doing cutting-edge work in their discipline. Topics ran the gamut of zoological research: biochemistry to paleontology, ecology to pathology, taxonomy, and evolution. The talks might not always touch on your area of interest, but they were fascinating. I didn't miss many.

Gary Nelson appeared on stage in mid-February: tall and gangly, wearing a blue and white plaid shirt with a string tie and flesh-colored glasses. An unlikely figure to set off fireworks at the end of his lecture. Gary was a research associate at the American Museum of Natural History in New York. He was the leader of a group of scientists promoting new approaches to evolution and biogeography: cladistics and vicariance, respectively. The primary focus of his presentation was to reintroduce the work of three scientists (Willi Hennig, Leon Croizat, and Daniele Rosa). Gary thought they had important things to say, and a reexamination of their ideas and methods, which he outlined, might be beneficial to the discipline. I judged his presentation as thought-provoking as well as entertaining.

Dick Fox, our paleontologist, was first on his feet when the moderator called for questions. Dick was a terrifying specter, particularly to students presenting their thesis work. He was the department curmudgeon and never hesitated to call speakers on the carpet if he found fault with their facts or reasoning. Dick looked like he would explode. I couldn't hear what he said. He was seething and spoke through clenched teeth. When Dick finished, he turned from his front-row seat, climbed the 30 steps from the well of the hall to the exit, and vanished.

I didn't realize Gary had been on campus for several days arguing with Dr. Fox. The eruption we witnessed was the culmination of heated debates between the two. Gary's presentation was the proverbial last straw. I only knew what I heard in the previous hour and thought, "Wow! Anything arousing such passion is worth looking at." I immediately ordered copies of Hennig's and Croizat's books and started reading. The upshot of Dr. Nelson's presentation made me determined to include sections on the evolution and biogeography of *Parelaphostrongylus* in my thesis.

The summer of 1976 was the summer of the snail. I knew the worm's identity: *Parelaphostrongylus odocoilei* from the west, not the meningeal worm from the east. The next step was to identify the mollusks transmitting the infection to deer in Jasper. I needed a lot of snails from different habitats if I wanted to learn how and where deer acquired the infection. From mid-May, when the snow began to melt, to mid-September, when it started falling, my week was the same. Monday, I drove the 225 miles to Jasper. Monday and Tuesday, I explored my collecting sites rolling over rocks and logs, searching for slugs and snails. Tuesday evening, I made the return trip. The remainder of the week was devoted to processing my catch. I identified each mollusk, not a simple task, at least at the beginning. Each snail was minced, artificially digested, and the nematodes present counted. Most days started at 7 am and ran for 12 to 16 hours. That summer, I examined over 8,500 animals belonging to 15 species, and 122 individuals (or 1.4%) harbored nematodes. The two prime suspects critical to transmitting the infection to deer were the small, black slug Roy Anderson fingered as the intermediate host of *Parelaphostrongylus tenuis, Deroceras laeve*, and *Euconulus fulvus*, a tiny, conical snail measuring about 2 mm in diameter.

I tried to avoid encounters with tourists while collecting snails. What I was doing didn't make sense to the lay public, and I found it best not to try to explain myself. I continued to collect deer feces when the opportunity arose. One day, I spotted several deer foraging in a small stand of lodgepole pine and aspen between the Athabasca River and the highway bordering Jasper's southern edge. I focused my attention on a young female hoping she would drop some pellets, so I could go back to snail hunting. Deer are like mice. They poop all the time. My frustration grew as a few minutes morphed into half an hour with no action. She broke cover for the open area beside the road. Cars stopped, and tourists piled out to admire nature and take pictures. Still hidden, I focused on her tail. As soon as the camera crowd started offering food, her tail went up, and poop dropped to the ground.

I dashed out to collect the Milk Dud-like offerings. The first reaction of the tourists was annoyance because I was spoiling their photo-op. Irritation turned to alarm when I dropped to my knees and started scooping the poop into a plastic bag with my bare hands. I apologized and walked back into the woods. I heard mumbled concern about my sanity as I disappeared into the forest. It wasn't the first, or last, time fretful visitors reported me to the rangers. I let them use their position to assure the public I was harmless.

Fall was devoted to laboratory studies of how the larvae in deer poop infect their snail host. I obtained specimens of a land snail (*Triodopsis multilineata*) from Dave Ashely, a BGSU alum, then engaged in his PhD work at the University of Nebraska. It took time, but I learned how to care for them and get them to reproduce in the lab. I exposed snails to larval nematodes on filter paper. I killed snails at regular intervals, embedded them in paraffin, cut thin sections, and put them on microscope slides (all of which I learned courtesy of Dr. Berg at Hiram). I stained the slides and followed the penetration of the worms into the foot of the snail, tracking their development. The process from penetration to the stage infective to deer took less than three weeks. With the worm's identity confirmed, the snail hosts in Jasper identified, and the parasite's development in the snail locked down, I could concentrate on the evolutionary and biogeographic questions raised by the seminar conflict between Gary Nelson and George Fox.

During the fall and winter (1976–77), I spent hours every day in the library reading and taking notes. There were times when I needed a break from deer, plate tectonics, and worms. I needed to read something for fun. The University of Alberta boasted over a dozen libraries. Most, like the Science Library, were devoted to specific disciplines. The Main Library was a refuge full of novels offering a brief respite from science. I fell in love with Mark Twain, or at least his writing, at an early age. Tom Sawyer and Huck Finn were favorites, but I didn't know much about the rest of his work.

I headed for the 'Main' in search of something fun. I found the Twain section, which included a row of books with identical binding and the heading "The Complete Works of Mark Twain," followed by the title of the specific volume. I haphazardly pulled a book from the shelf and read the title page. At the bottom was the following, "The first volume of the first 250 sets are signed by the author." I headed to my left and found the first volume, *Innocents Abroad*, opened the cover, and there it was, Mark Twain/Samuel Clemens written in fountain pen ink. There was nothing else on the page. I checked it out and read it over the next week. Before returning the book, I had a larcenous thought. It would be so easy to take a razor blade and excise the page. Since I was the first person to check the book out in over a decade, nobody would know, at least until I graduated and was gone. I couldn't do it. I took the book to the circulation desk and suggested to the student worker the volume was valuable and perhaps should be in the Rare Book Room or some other protected location. She glanced at the signature, sighed, shrugged her shoulders, and tossed it under the counter. I was not hopeful; however, on my return to the 'Main' several weeks later, *Innocents Abroad* was not the first volume in the Complete Works of Mark Twain.

With the laboratory work and library research completed, I had everything needed to write my thesis. I had to sit down and put pen to paper. I applied for and received a Dissertation Fellowship — the only one offered in the department. I would be released from any teaching duties and paid to write! One of the single offices was available, and I grabbed it. Time and solitude were what I needed, and I had both.

Making yourself known in the discipline is critical to advancement in science: publishing and presenting. I had two publications from my

Master's days at BGSU.[1,2] Al Bush, who returned from his self-imposed exile, and I published the description of a new species of nematode from a captive Ball Python,[6] and I got a short note accepted describing the unique occurrence of an acanthocephalan in a mouse.[3] I still hadn't presented a paper at a scientific conference. My first foray in oral presentations occurred at PUBS (Prairie Union Biological Seminar), a meeting for students only, scheduled at the University of Lethbridge in southern Alberta. I rehearsed and rehearsed and rehearsed. The members of the parasitology group critiqued my effort. I was ready but scared shitless when I was introduced and took the stage in early February 1977. My speech pattern mimicked the hucksters who read the fine print in ads at speeds no one can comprehend. Then I had an out-of-body experience. My disembodied self said, "Take a deep breath and slow down, or you will pass out." I did, and all was well.

The annual meeting of the American Society of Parasitologists was held in Las Vegas later in the summer. I hadn't attended the national conference since the 1973 meeting in Toronto to meet Dr. Holmes. Everywhere from Edmonton was expensive, and I didn't have anything to say anyhow. Now I did. It was time for my national/international debut. Conference organizers divide presentations into blocks on similar topics, with several sessions running concurrently. To avoid conflicts, the subjects of concurrent sessions typically featured unrelated topics: say biochemistry and taxonomy. Most biochemists don't care about taxonomy and vice versa. I controlled my nerves, focused on the material, and spoke clearly, a far cry from my first attempt at PUBS. The talk went off without a hitch; however, since few people were familiar with the subject of my work, I only received a smattering of questions. On the bright side, nobody attacked my results or interpretation of the data.

Kathy went back to Ohio while I was in Las Vegas. I returned before she did, so I hustled out to the airport to pick her up when her flight landed. I could tell something was up when I met her at the gate. We had been trying to get pregnant, so I had a good idea of what was behind the Mona Lisa smile when she asked, "Do you notice anything different?" I picked up her bags and replied, "You're pregnant." I should have jumped up and down or done something more demonstrative. I was happy. I was thrilled. I should have done a better job of showing it. It took a while, but I finally convinced her how delighted I was.

The writing was progressing well. I write instinctively. Whatever I learned about grammar in elementary and high school was long forgotten. I write what sounds right. There is a style of writing science, but it is not stylish. Simple, declarative sentences (noun, verb, object) dominate, a few starting with prepositional phrases thrown in to break up the monotony. Bill Samuel made up for deficiencies in my prose. Bill was a ruthless editor and good at it. He helped in ways large and small to clarify ideas and tighten arguments. Progress was steady, and Bill approved.

I had no doubt I would finish writing by Christmas and asked Bill about scheduling my thesis defense early in 1978. The upshot of this had other, long-range implications. I would have to start looking for a job. My original goal for completing the PhD, "to see if I could do it," was in sight, and now I had to put it to some practical use. I devoured the employment sections of *Science* and *The Chronicle of Higher Education* every week. I wrote letters, hundreds of letters, inquiring about positions north and south of the border.

As the semester wound down and my defense approached, I had to choose an external examiner for my committee. Roy Anderson, the world's expert on *Parelaphostrongylus*, was the obvious choice. While he and Bill Samuel had a cordial relationship, Roy seemed annoyed we were encroaching on his territory. Bill nixed him from consideration. We settled on Ralph Lichtenfels, Curator of the National Parasite Collection in Beltsville, Maryland. Ralph was an imposing figure with elegantly coiffed silver hair and an immaculately trimmed matching beard and mustache. While Ralph looked more like a bank president or plantation owner from a bygone era, he was an expert on nematode anatomy and taxonomy. Ralph accepted, and Bill scheduled my defense for February 2nd, 1978.

Christmas came and went. I finished the final draft of my thesis and distributed it to the members of my committee: Bill Samuel (Chair), John Holmes, Andy Spencer (physiologist and regular squash opponent), Jan Murie (mammalogist), George Ball (entomologist and taxonomy consultant) and Ralph Lichtenfels. Bill and I met regularly to game likely questions and potential problem areas. I read and reread the literature and my thesis to ensure I had everything down cold. Bill offered the usual advice meant to reduce my anxiety: nobody in the room knows more about this than you, and failures at this point were exceedingly rare. I started wondering if regularly beating Andy Spencer in our weekly squash games might have been somewhat impolitic.

At approximately 1 pm, I entered the department conference room, acknowledged the six men seated at the table, and parked my butt in the only empty chair. The next three hours were a blur. Most questions were straightforward and answered accordingly. George asked me who I consulted about the identification of the snails I collected in Jasper. I said, nobody. I read the literature and was confident my identifications were correct. George was a rugged individualist and liked to solve problems on his own. My response seemed to please him no end. George also questioned my use of "sacrificed" when I wrote about killing an animal. He asked, with the corners of his mouth upturned in an impish grin, if I carried out these actions during a full moon, or, perhaps, in costume? The question was rhetorical. George boomed, "Have enough respect for your animals to say what you did. You killed them!" I excised "sacrifice" in all its forms from the final draft.

After the last question, I stepped outside while the committee deliberated. 15 nerve-wracking minutes later, the door opened, and they invited me back in. Their smiles spoke volumes. Each man, in turn, shook my hand and congratulated me on an excellent performance and research project. They each paused to sign the title page of my thesis and acknowledge my successful defense. We retired to the Faculty Club, a short walk from the Zoology building, for a celebratory drink. When I got home, Kathy and I celebrated with a few tears and dinner at the most expensive restaurant we could get in on short notice.

The last requirement was the departmental seminar. I would star in one of those Friday afternoon gatherings. A 50-minute presentation, followed by questions, was required for graduation. The earliest date available was March 31st. While I was no longer a presentation virgin, my previous efforts topped out at ten minutes. It was back to work. I had to organize what I was going to say, make new slides, and rehearse, rehearse, rehearse!

I watched the auditorium gradually fill with students and faculty while sitting in a chair on the stage. Dick Fox, of stomp-out-the-room fame, walked in and took his usual seat in the front row. He still scared me, and for good reason. My assessment of deer evolution strayed into his area of paleontology. I also included a cladistic analysis of *Parelaphostrongylus odocoilei* and its relatives in my thesis and presentation. Dick was not a fan. 50 minutes passed in the blink of an eye. The applause, while not deafening, seemed more than polite. The questions started. After the thesis defense, I was well prepared. I kept glancing at Dr. Fox, waiting for an attack that never materialized. Bill and I retired to his office for a debriefing of my performance. I expressed my relief Dr. Fox had remained silent. Bill said he talked to Dick on his way back upstairs. Dr. Fox's assessment was grudgingly positive, and he observed, "At least he included fossils." As close to a compliment as I could expect.

Education is where you find it: the classroom, a seminar, a hallway conversation, or pouring through volumes in the library. I drew on all of them during my time in Alberta. But the most surprising was the 4th-floor coffee room. Coffee breaks were akin to the United Nations. Students and faculty at the U of A hailed from the four corners of the globe: Malaysia, Australia, New Zealand, India, the United Kingdom, several countries in Africa, Canada, and the United States. Exposure to the cultural heritage afforded in casual conversation over a cup of coffee or tea and a donut was life-altering. Before moving to Edmonton, I had rarely been out of Ohio: summer vacations fishing in Ontario as a child and a bus tour of Washington, D.C., with my mother during junior high. At Hiram, we had one international student from Nigeria. Within the confines of a few hundred square feet, I encountered the world. Discussion of politics and religion was fair game but not as important as our interest in understanding the workings of the natural world following the dictates of observation, experimentation, and logic.

I experienced racism from a quasi-personal perspective. When the provincial legislature proposed a tuition hike at UA, the undergraduate students were apoplectic. The rallying cry was to charge foreign students more and Canadians less. The target of the undergrad ire was the Chinese. White Canadians failed to appreciate that many of the Chinese undergrads at the time were the descendants of laborers who built the Trans-Canada railroad in the late 1800s. Nobody ever pointed a finger at me, an American, because I looked like them. The lessons I took from those experiences weren't captured in notebooks to be reviewed and committed to memory and regurgitated on an exam. They opened my mind and heart to the idea that we are more alike than different and want the same things for our families and from life. I learned how to do science, appreciate the diversity inherent in the human condition, and reject the ugly face of racism.

Nothing was happening on the job front. Then the phone rang. I got a call to interview for a position at the University of Richmond in Virginia. These visits follow a standard form: interviews with the Dean and faculty, presentation of a seminar, and dinner with faculty from the Biology Department. Then go home and wait. Kathy was closing in on her 8th month of pregnancy, and we had to sell the house. Kathy chose to take maternity leave to finish out her career as a physical education teacher, cheerleading advisor, track coach, and camping and cooking guru at Hillcrest Junior High. She felt a celebration was in order and booked a week in Hawaii. Later, we both marveled a woman eight months pregnant was allowed to get on a plane.

Our realtor suggested combining the trip with selling the house. We cleaned, decluttered, and took off for paradise. On our first full day, we rented a car and drove the perimeter of the Big Island. We toured volcanos, black beaches, and macadamia nut farms. When we checked back at the desk to get our key, the manager handed us a pile of papers — phone messages from our agent. We called and learned we had had five offers on our home the first day — each better than the last. We listed the house for $55,000. The last offer was $58,000 plus closing costs, plus (a big plus) we could stay for a week after the closing on June 1st. Driving around the island more than paid for our trip. The first offer was fair but not close to the final one. We spent a little under $24,000 for the place four years earlier. We did invest some cash and sweat equity in it, but not enough to justify doubling our money. The tar sands and a plunging vacancy rate deserved most of the credit.

Contractions began on May 1st, and Kathy went to the hospital the following day. After 44 hours of labor, our first son, David, was born three minutes after midnight on May 3rd. Kathy spent four days in the hospital recuperating from a C-section. The day after she and David came home, the phone rang, and the folks in Richmond offered me a job. I became a father and a newly employed academic all in the space of a week. We had a month to pack, attend graduation, and say goodbye to Canada.

On June 7th, I drove Kathy and five-week old David to the airport for the flight home. Shunda and I headed east in our Nissan Sentra. Four days and two thousand miles later, I pulled into Kathy's parents' driveway at 1321 High Street, Louisville, Ohio. Our Canadian adventure was over. Our marriage survived graduate school. We were still in love, new parents — and I had a job!

8. The University of Richmond

If there's a single lesson life teaches us, it's that wishing doesn't make it so.
— Lev Grossman, The Magicians

Richmond, in 1978, was a small town. The invasion of corporate headquarters and the city's transformation into a business and financial center were still a decade off. Richmond was dominated by families who traced their lineages back ten generations (with slave owners in their pedigree) and wore jodhpurs while riding on the weekends. When Kathy graduated from high school, she and her Aunt Grace stopped in Richmond on their way to visit family friends in Alabama. During breakfast at a local diner, their waitress asked where were they headed? Aunt Grace replied, "Farther south." The waitress replied, "Honey, you can't get any farther south." Her meaning was clear. Richmond was conservative to the core. We were in a foreign country for the first time in our lives.

Our new house was a hair inside the city limits and three blocks from Monument Avenue, a leafy boulevard lined with statues of the heroes of the War of Northern Aggression. We noticed the paucity of children. We learned once the city schools desegregated in the early 1970's white flight achieved supersonic speed. The only children in our neighborhood were two preschoolers, but their parents planned to send them to a private (i.e., all white) school. The grandmother of those children shared with Kathy, she "would give up all those new-fangled appliances if she could get a black back in the kitchen." — conservative and racist. Ours, we were thinking about one more, would attend public school.

I was the youngest member of the department by two decades, with one exception, and the only one untenured. They all had weathered the transition of the University from near bankruptcy to prosperity and a healthy endowment. Most were racist in the sense they would deny it if confronted but exposed the underlying truth in casual conversation and the jokes they told. Despite their obvious shortcomings, I liked them. I wasn't going to change them, and I needed their support if I was going to work my way through the thicket of academic politics and keep my job. I walked a fine line to maintain my integrity while not overtly calling them out. The exception was Dave Towle. Dave was a New Hampshire transplant and former Peace Corps volunteer. He and his wife, Betty, would become, and remain, friends long past Richmond.

We made friends among the younger faculty at UR, socialized with them occasionally, and attended the biology department's required functions. Then serendipity intervened. Right around the time David turned one, Kathy raised the issue of his moral and ethical development and the possibility of attending church. I wasn't interested. She persisted. I finally agreed but said I would only consider a Unitarian-Universalist congregation, the adopted church of my youth. A church I had abandoned by declaring myself an atheist at the age of 14. Nobody bothered to tell me atheists were welcome and not uncommon in UU congregations. Finding the church marked a turning point in our lives. The Reverend David MacPherson, a true New England liberal, led the Richmond enclave. His congregation was an oasis in a desert of southern conservatism. Dave and Betty Towle joined us. Our friendship deepened when our sons, Reid and Aaron, were born a few weeks apart in 1981.

We found a spiritual home and could fend off repeated invitations to attend church from our neighbors by saying, "Thank you, but we are happy with ours." The UU church became the hub of our social and political lives. I joined the Worship Arts Committee, planning and presenting lay-led services. Kathy started a chapter of the Homemakers for Equal Rights Association (HERA), joining marches supporting the Equal Rights Amendment. She was the go-to person when the local newspapers needed a photogenic face for stories about the Women's Movement. Our pictures graced the pages of the Richmond Times-Dispatch and Richmond News-Leader on more than one occasion.

In the late 1960s and early '70s, UR was a struggling institution with a real possibility of closing its doors due to its meager endowment. And then it happened. E. Claiborne Robins, CEO of A. H. Robins Pharmaceuticals, wrote the school a check for $50 million. When the young Robins was a student in the 1920s, he came close to dropping out due to money issues. The school awarded him a scholarship of $250 that allowed him to complete his degree. He returned the favor. UR purged the Board of Trustees of the 'no dancing allowed on campus' Baptists and replaced them with corporate types. The Robins' family chipped in another $100 million, and more followed from various sources. The new UR had visions of becoming the Harvard, or Princeton, of the South. They expected new hires (me) to carry the same teaching load as the established faculty, do research, and get grants.

I became embroiled in the evolution-creation debate raging in the early 1980s. Liberty Baptist College, now Liberty University, headed by Jerry Falwell, was attempting to transition from a sleepy Bible college to a recognized institution of higher learning. As a biologist, I strongly felt evolution was the discipline's foundational concept and should be defended vigorously. Liberty applied to have their biology program certified by the state to train teachers in the subject. I learned that the college earned accreditation from the Transnational Association of Christian Colleges and Schools (TRACS), an organization whose members accepted and promoted biblical inerrancy. I circulated the information to the State Board of Education and stopped the process cold, pending a full review of their program. After reading the requirements for certification, I knew I was only delaying the inevitable. Liberty promised to keep biblical teaching in the Religion Department, and teacher training in biology began a year later.

My first set of teaching evaluations nearly killed me! I ranked well below the 50th percentile. On closer inspection, I realized the scores were presented using a normative scale. For example, take 100 faculty scored on a 5-point scale, and your score of 4.2/5 puts you in 90th place — the bottom 10%! My scores were middle of the road numerically but terrible by rank. There was no question I needed to improve.

Colleges and universities determine tenure by contributions to the institution in three areas: teaching, research, and service. There is, however, no objective standard for any of them. In the end, the decision relies on the goodwill of your colleagues and whoever has to give their approval above the department level. It can either be a rubber stamp or a crapshoot. I worked diligently on my lectures and classroom presence. The student evaluation form was revised to separate classroom performance from popularity and eliminate the unfairness of normative ranking. That helped. I was engaged in research and publishing, leaving service. I managed appointments to several committees at both the departmental and university level. I can't say I distinguished myself, but I was contributing. My outlook improved when feedback from my chair and other members of the department was overwhelmingly positive.

The university was, as typical in the 1800s, initially an all-male institution — Richmond College. When they finally enrolled women, complete integration of the genders was a step too far. A co-ordinate women's school, Westhampton College, was located on the opposite side of Westhampton Lake. The path connecting the two schools had a gate, locked in the evening to curtail any extra-curricular fraternization. Legend has it some of the young men became excellent swimmers in furtherance of their hormonal drive and the encouragement of young ladies waiting on the opposite shore.

The two colleges shared science and math faculties but had separate departments in the Arts and Humanities. Men graduated from Richmond College and women from Westhampton. By the time I arrived, all departments had merged, but, in a nod to history, advisors served one college or the other, men or women, not both. I volunteered to be an advisor and drew the young women of Westhampton. Most faculty avoid advising like the plague. I loved it. I enjoyed meeting students individually, learning about their hopes and dreams, and suggesting courses to best achieve their goals. I like to think I clarified a sometimes confusing process and helped them select classes matching their interests and aims, rather than merely checking a box.

"We" got pregnant during the summer of 1980 with an expected due date in late April. On Friday afternoon, April 17th, I received a call from the Henrico County Hospital. Kathy was in labor, and my presence was urgently requested. That was a surprise. Kathy had a C-section with our first son, and her current doctor would not countenance a vaginal delivery

for the second. The birth of our second child was scheduled for the following week. He had other ideas. I dropped everything and got to the hospital
a few minutes before 5 pm only to hear Kathy pleading for the delivery to
be delayed until the 18th, her father's birthday. The doctor smiled, shook
his head, and said: "This is happening now." A nurse whisked Kathy to the
delivery room, and I donned a set of scrubs as my entre to the birthing
theater.

Once in the delivery room, I was told my job was to support and comfort my wife. A small tent straddled Kathy's abdomen, blocking my view of
the exciting stuff. Initially, I obeyed the rules. I held her hand and offered
words of encouragement and support. The doctor said, "Let's take care of
this scar." Her first delivery employed a vertical incision resulting in a
prominent scar about 8 inches long. I heard a loud bang as the excised tissue landed in a pan on the other side of the divide. Curiosity got the better
of me, and I rose slightly off my stool to get a better view of what was going
on. It only took a minute for the physician to notice my lack of attention
to my wife and scold me with, "Your job is on that side; this side is mine."
Sufficiently chastised, I returned to my assigned role. Reid Fenton Platt
made his debut at 5:31 pm. Mother and son were fine, and we were a family of four.

Fast forward two years. Our kids were growing. David, the elder,
attended pre-school at the Jewish Community Center for two years and
was ready to start kindergarten. Our neighborhood school was integrated;
however, as a result of white flight, it was primarily African-American —
both students and teachers.

One of David's classmates was Jennifer Robb, daughter of Lynda Bird
Johnson and Chuck Robb. Lynda Bird was the daughter of former president Lyndon Johnson, and Chuck was the governor of the Commonwealth.
Chuck had higher political aspirations, and sending his daughter to private
school might have seemed elitist. David and Jennifer became fast friends.

Kathy arranged play dates, and Jennifer arrived at our home under the
watchful eye of a couple of State Troopers. I made occasional visits downtown to pick up David from an afternoon of scampering around the
Governor's mansion. I chatted with Chuck on occasion while his staff
gathered David's things. Jennifer attended at least two of David's birthday
parties. Lynda Bird sat dutifully in our living room chatting with the other

mothers while the kids played in the backyard, ate cake, and David opened presents. Richmond was a small town.

My research was a scattershot of projects. I published six papers from my PhD,[4,5,7,14-16] three with my students[9,10,17] and two on unrelated subjects.[12,13] The reason? I was trying to be something I wasn't. I decided to go back to my Master's work and study turtles.[11] Westhampton Lake, right in the middle of campus, was full of them, and catching turtles did not require a collecting permit, a significant plus. I trapped and dissected close to 100 turtles in my time at UR, but I didn't know what to do with the data. I tried to become an ecological parasitologist, but math was my downfall. It never made sense to me.

Tenure is wildly misportrayed in film and on television. Typically, someone in the humanities is trying to finish a book. If successful, they get tenure. If not, they drive a taxi or tend bar for the rest of their life. The reality is much more of a slog and based on a combination of factors. The candidate, generally in their sixth year, compiles an exhaustive account of their activities in teaching, research, and service, providing extensive documentation of their accomplishments: courses taught with enrollment numbers, syllabi, teaching evaluations, and even testimonials from former students; publications, grant applications, and a statement of future research goals; committee service, your role, and accomplishments. Finally, the candidate requests letters of recommendation from colleagues in other departments and experts in their field from other institutions.

At UR, the department assessed the dossier and made a unified recommendation to the Dean. Then it went Dean to Provost, Provost to President, and finally from the President to the Board of Trustees. The Board and only the Board has the authority to grant tenure. If the department makes a strong recommendation for or against tenure, the remaining decisions should be a formality. However, and this is a big however, it is not always the case. The Dean, Provost, President, or the Board can do anything they damn well please.

Gresham Riley was Dean when I interviewed for the position. I never spoke with him in the following four years before he departed to become President of Colorado College. During my initial interview, Gresham asked few questions while spinning the massive Yale class ring on his right hand. I have no idea if my case would have had a different outcome if he had stayed. The Provost was another matter. Mel Vulgamore was a friend, and Kathy and Mel's wife, Nan, developed a more than cordial relationship. His departure the year before I came up for tenure to become President of Albion College in Michigan was disconcerting.

Dean Riley was replaced by Sheldon Wettack, who was described years later by a friend who was attending a meeting of college administrators as "straight out of central casting for a college president." I tried to get to know Dean Wettack in the year before my tenure application. My efforts were either unwelcome, patently transparent, or both. We never connected. Zeddie Bowen replaced Mel as Provost a year later. I never got to know much about him other than he seemed like a decent person on the few occasions we were in the same room together.

The department unanimously recommended me for tenure which should have sealed the deal: rubber stamp, rubber stamp, rubber stamp, rubber stamp, and I have a job for life. How wrong I was. Early in the spring semester, I received a call informing me I was to meet with Dean Wettack at 3 pm on Friday. I was confused and a tad concerned, but I told myself the department recommendation would carry the day. As soon as I entered the room, I knew it wasn't going to be good. Dean Wettack started blathering on about something I couldn't quite decipher. It was all white noise until I heard, "You aren't the kind of person we want to invest in for the next 30 years." I tried to counter and argue my case, but even an idiot, a hurt, wounded idiot, could see it wasn't going anywhere. I walked out of his office, down the hill to the Science Hall, collected my things, and drove home to tell Kathy the news. I was 36 years old, married with two children, a mortgage, and my life had been turned upside down. For one of the few times since childhood, I started to cry.

Kathy was, in turn, devastated and furious. She started plotting a strategy for my appeal. We put the best face on the situation we could, had dinner, and tucked in the kids for the night. The next day I talked with Dave Towle on the phone. He was stunned, as were other members of the department. They made a few calls but to no avail. I met with the Provost, but I could tell I wasn't worth getting in a fight over with the Dean. They were both new and had to work together, so 'hasta la vista' Tom. If Mel Vulgamore had remained Provost for one more year, things might have been different. He was a friend and had the institutional heft to put

Sheldon in his place, but who knows? The President was an empty suit, and there was no point trying one more step up the administrative ladder. My career at the University of Richmond was over.

For the first time since I decided to take academics seriously during my junior year at Hiram, I failed. The problem was I didn't know why I failed. "You are not the kind of person we want to invest in for the next 30 years." is an excuse, not an explanation. Was it the lack of a coherent research program? Lack of grant money? My involvement in the evolution/creation debate? Dean Wettack came from an evangelical Christian institution, and that may have ruffled his feathers. I didn't know why then, and I still don't.

During my final year, I became a ghost. Old friends and colleagues were no longer available for lunch. They developed an enhanced search image allowing them to find a door, a hallway, or somebody (anybody) to engage in serious conversation if the soon-to-be ex-colleague appeared on their radar. It was akin to the plague; I had it, and they didn't want to catch it. The stink of failure followed me everywhere.

I applied for every academic job I could find, but they were few and far between. I was offered a job at an HBCU (Historically Black Colleges and University) near Richmond. The position included teaching Biochemistry and Microbiology, courses I had no qualifications to undertake. I declined. Their students deserved better. I was also offered a job as Assistant Dean for Advising at one of the colleges comprising Rutgers University. I knew as soon as the plane landed in Newark, I didn't want the job, but I wanted them to offer it to me. I worked hard to impress. I liked the Dean, who would have been my boss, and he liked me. They offered, and I declined. I never felt better saying 'No' to a job I didn't want, in a place I didn't want to be.

I don't recall how I discovered the Career Opportunity Institute at the University of Virginia. COI was the brainchild of Larry Simpson, Director of Career Planning and Placement at Jefferson's university. The goal was to prepare academics (failed, disillusioned, or otherwise) to transition from an academic career to one in business. The six-week course included brief introductions to accounting, management, basic business practices, job-hunting, and interviewing. Before our first meeting, we took a battery of psychological and aptitude tests, including the Myers-Briggs Personality

Assessment. We were 32 strong from across the country, representing all disciplines, and shared one trait. 31 of 32 of us were introverts according to Myers-Briggs. Other than our introversion, we were a bag of cats. We received enough information in each content area to make us dangerous.

The culmination of the course was the 'informational interview.' We called someone, anyone, employed in a field of interest and asked if we could talk with them in person about their job. We were coached not to inquire about employment. We only wanted to find out more about their field. At the end of our intrusion into their lives, we asked: "Is there anyone else you know I might talk to?" Two things shocked me about this process: 1) how many people were happy to waste 15–20 minutes talking to a complete stranger, and 2) how many were willing to throw someone under the bus. The idea was if you did this enough times, somebody would hire you.

The initial calls were like asking a girl for a date in high school. First time — think about it and find something else to do. Next, pick up the phone and hang up without dialing. Then pick up the phone, dial part of the number and hang up. The fourth time dial the entire number and hang up. Eventually, you screwed up enough courage to let the call go through and state your business. I made over 100 of those calls, and no one ever said 'No!'. I also never had anyone fail to give up a friend (or someone they wanted to stick it to) when asked.

I received my last check from UR in July 1985 and applied for unemployment. I met weekly with my caseworker to update her on my attempts to secure employment. The one-year timeline for academic positions was mystifying to that career bureaucrat, but I was convincing, and the checks kept coming. Kathy was the salvation of our family. My wife was never one to sit on her hands and cry, "Woe is me." She started a business, *Distinctive Desserts*, out of our home. When she realized the abysmal profit margins in pies and cakes, she branched out to general catering. During my final year at UR, Kathy latched onto several regular customers for lunch, occasional dinner soirées, and other miscellaneous functions while taking care of two small children. I subbed in as a waiter/bartender when needed. Between my unemployment check and Kathy's catering, we made more money than I earned as an Assistant Professor. Go figure.

Informational interviewing was interesting but not terribly rewarding. As I sat in the office of a local printer waiting to talk to the head of HR,

I didn't know it, but the worm (pun intended) was about to turn. The company printed all manner of magazines, journals, and newspapers for distribution in the mid-Atlantic region. I have no idea why I was there other than somebody gave me a name, and when I called, he said: "Sure, I'd love to waste some time." On the wall opposite from where I was cooling my heels was a floor-to-ceiling rack with the current issues of all of the magazines the company printed. Directly in my line of sight was *Science*, a principal source of ads for academic jobs here and abroad. I hadn't perused those offerings in several months and thought: "What the heck!" In the middle of the first page of job placements were two ads, side-by-side. On the left, Midas Rex was looking for PhDs (any subject) to teach orthopedic and neurosurgeons how to use a high-speed pneumatic drill in surgery. On the right, Saint Mary's College, Notre Dame, IN, wanted someone to teach non-majors biology, Invertebrate Biology, and Freshwater Biology or Parasitology. Informational interviewing was about to work its magic, but not quite in the manner the folks at COI envisioned.

I made note of the issue and page number before I was called back for my interview. After a wasted half-hour, I headed straight for the nearest public library, found the issue of *Science,* and photocopied the page with the ad for Midas Rex. I carefully read the requirements, made copies of the requested documents, and sent them off with a nicely composed cover letter. Several weeks later, the folks at Midas Rex called and asked if I could come to Fort Worth, Texas, for an interview. The trip was a one-day affair, and they would pay me $100. I said I would be delighted, and off I went.

Midas Rex operated from an industrial park on the outskirts of the city. They showed me the manufacturing facility and gave me a brief overview of the position. I would travel around the country, teaching orthopedic and neurosurgeons how to use their product in a classroom setting. The drill was no bigger than your thumb, spun at about 80,000 rpm, and could cut bone and metal like the proverbial hot knife through butter. I was interested. They gave me a check for $100 and put me on a plane back to Richmond with the assurance they would be in touch.

While waiting for a call from Texas, I pulled out the photocopied page and showed Kathy the ad for Saint Mary's. She read it carefully, at least twice, and told me in no uncertain terms: "If you don't apply for this, you will kick yourself for the rest of your life." I applied.

Midas Rex called and invited me to attend one of their classes. My job was to observe, help out wherever I felt comfortable, and they would pay me $100/day for four days. I had no idea what to do, but there seemed to be a lot of cleaning and grunt work needed, so I jumped in and did what I could while observing the structure and presentation of the class. The student-instructor ratio was 4:1. Instructors took turns narrating videotapes showing the students various applications of the high-speed drill, which the students repeated on bones from livestock. The activities increased in complexity as the course wore on.

When the class concluded, I helped with the cleanup, was handed a check for $400, and told they would be in touch. They called again and invited me to another course, same length, same pay. This time I would take the class with the physicians who were forking out $1500 for the privilege. Again, I helped with the setup and cleanup, but in between, I watched, listened, and cut bone and metal. I received another $400 and waited.

In mid-December, I became an instructor for the Midas Rex Psychomotor Institue for $40,000/year, just a tad less than twice what I was making at UR. I flew to Fort Worth for training in mid-January. In the middle of the first week of training, Kathy called with the news Saint Mary's requested a telephone interview for their position. When it rains, it pours! I have to admit to a certain amount of guilt doing a phone interview for a job from a hotel room paid for by my current employer.

The conversation went well, and Saint Mary's invited me to campus for the standard dog-and-pony show: meetings with students, faculty, administrators, and a lecture. Doris Watt, head of the search committee, met me at the airport in early February. Doris is an ornithologist, and as I quickly learned, a bit quirky. She drove an aging Nissan with a malfunctioning defroster. It was about 10°F outside, pitch black, with mounds of snow lining the streets. Doris thought a tour of the city and the Notre Dame campus, right across the street from Saint Mary's, would be entertaining and informative. At one juncture, she pointed into the void and said, "If it wasn't so dark, you could see the Golden Dome right over there." as I tried to follow her finger while scraping frost from the inside of the window with a credit card.

The next day I met with faculty, staff, students, and administrators of all stripes. I answered their questions thoroughly and honestly. However, I could not tell them why I had been turfed at UR because I didn't know.

I scheduled this clandestine meeting between my duties with Midas Rex. I went back to my paying job, which took me to Key West, Park City (Utah), Dallas, and New Orleans. Saint Mary's called and offered me the position. After the uncertainty of the last year and a half, I had options. The choice was lots of travel and $40,000/year plus bonuses versus being at home doing a job I loved for $22,000/year and not a snowball's chance in hell of ever seeing a bonus check. Oh, there was the niggling thought that being denied tenure once, it could happen again. I don't like traveling, but the $18,000 differential between the positions was a lot of money. My heart won out. I told Saint Mary's I would start in August 1986. I had to figure out how to continue with Midas Rex until I began receiving a paycheck from Saint Mary's. I still had a family to support.

The message from COI was "to always play your cards close to the vest." If you were going to take a new job and were required to give two weeks' notice at your old one, tell your boss 14 days out, not a moment sooner. While I didn't love my job with Midas Rex, I didn't hate it either. The people in the company loved me and treated me well. I could not, in good conscience, stab them in the back. I made my decision in early March and told my boss I would be leaving in June. I added I would like to continue working, if at all possible. She was disappointed but said she could certainly keep me busy until my departure. She did. I left on good terms and even had a bonus check in my pocket when I walked out the door for the last time.

9. Saint Mary's College

> *Though nobody can go back and make a new beginning… Anyone can start over and make a new ending.*
>
> — Chico Xavier

Moving and beginning a new job is always stressful. I would have to meet new colleagues and students, learn the institutional ethos and culture, organize my classes, and develop research programs for my students and myself. The only thing I knew about South Bend was that it was the home of the University of Notre Dame Fightin' Irish. I wasn't a fan. My hometown was a football town. If you went to public school, you rooted for Ohio State and hated Notre Dame, especially if your father and brothers were Buckeyes. If you attended parochial school, the reverse was true. My second favorite team was whoever was playing Notre Dame that weekend.

Saint Mary's College was founded in 1844 by the Sisters of the Holy Cross to educate young women, a complement to the all-male University of Notre Dame, located just across the street. An attempt to merge the two schools in the late 1960s failed, and Notre Dame opened its campus to women. Saint Mary's maintained its status as one of a dwindling number of single-sex colleges despite their often rocky relationship with the larger and more prestigious institution across the way.

Are there differences teaching at an all-female college compared to a coeducational institution? In the sciences, I would say no. Our students wanted to be doctors, dentists, and scholars and required the same training as any student. Things might be different in the humanities, but that is beyond my ken.

Are there differences between coeducational and all women's schools? Without doubt. The absence of men in the classroom allowed our students the freedom to be themselves without the pressure to impress their male counterparts. My students appeared in class sans makeup, wearing pajama bottoms and fuzzy slippers, with their hair still wet from a quick shower before dashing to class. In an all-female classroom, the women do not defer to men when asking or answering questions. There are no men to take over in the laboratory, leaving the women as spectators. If I had a daughter, I would strongly encourage her to consider an all-women's college.

The six years I spent advising young women at Richmond stood me in good stead when I took up my position at Saint Mary's. When faced with bad news, men lie or get angry. Women lie, get angry, or cry. I learned early on to have a box of tissues on my desk when the tears came. The best strategy is to hand the box to the student and wait. Don't try to comfort her; just patiently allow nature to take its course and continue where you left off when the lacrimal glands run dry.

Film and literature frequently use the trope of romantic entanglements between a "hot" older professor and a comely young undergraduate as a device to create tension in their tales of life in higher education. Occasionally a new acquaintance (male), upon learning of my situation, would, in a conspiratorial tone, intimate that I must be having the time of my life. The notion of sex with my students was never spoken but clearly implied — wink, wink. While I am sure these liaisons occur, I never heard of any during my 28 years at the college.

When I arrived at Saint Mary's, I was 37 and a generation older than my oldest students. While many of them were attractive, emotionally, they were children. The idea of a physical relationship with any of them was, for want of a better word, icky. On a more practical level, I could never understand why a faculty member would hand that much power to anyone, let alone an unpredictable teenager. Once you have sex with a student, your personal and professional life is in their hands. One word from your unhappy paramour to the Department Chair or Dean will likely end your career.

Was I ever confronted by a student batting her eyes, proclaiming she would do "anything" to raise her grade? Another cringe-worthy trope of bad Hollywood films. Possibly, but if it happened, the approach was not overt, and I probably responded by suggesting she study more. My wife claims I don't know how to flirt. I have one example suggesting she may be right.

During my short career with Midas Rex, we traveled to Key West, Florida, to teach the use of our high-speed drill to neurosurgeons. It was late January or early February, and while warmer than the northern Midwest, it was not swimsuit weather. However, I was determined to get at least a bit of a tan. One afternoon, when we finished for the day, I headed for the beach, commandeered a chaise-lounge, and laid in the sun with temperatures hovering in the low 70s and a light breeze. I started to nod off when I heard a woman's voice issue an unusually cheerful "Hi!" I looked up and responded with a tentative "Hello." That brief encounter unfolded as follows.

"Didn't I meet you in the bar last night?" she inquired.

"I don't think so," I replied.

"Are you sure? You look very familiar," she continued.

"I've only been here for two days, and I haven't been in the bar."

"Oh," she mumbled as she turned and walked away.

My wife is right. It isn't easy to flirt if you can't recognize the opening gambit.

What I found at Saint Mary's were terrific colleagues in the Department of Biology. Students who were (on the whole) bright, dedicated and reliable. Oh, and a home for the next three decades.

We purchased a house on the south side of town. It was large and in our price range. By South Bend standards, it was far out, 15 minutes from campus, but there were lots of space for the kids to play and lots of kids their age. I took a 50% pay cut to return to academic life. The lessons of history weighed on me as I prepared to take up my position at Saint Mary's. There were no guarantees when it came to tenure. I learned at Richmond if "they" (colleagues or administrators) want to get rid of you — they could, and you had little recourse. I made two promises. First, I would not try to be something I wasn't. My research would focus on the things I enjoyed and did well. Second, I would do the job the way I felt was sustainable for a career lasting decades. If it wasn't good enough, perhaps I should consider a different occupation.

My teaching duties were similar to those at Richmond: Non-majors Biology, Invertebrate Zoology, and Parasitology. My charge was to develop a rigorous but stimulating non-majors class for students taking biology to fulfill their science requirement (i.e., students who would rather be somewhere else). My predecessors were judged either too easy or too disorganized. My job was to present a course that made sense and might prove useful later in life. It was supposed to be challenging but not impossible.

Many faculty do everything in their power to avoid introductory courses, especially those for non-majors. I loved the class and never thought once about trying to palm it off. I worked and reworked my lectures relentlessly over the years. I read widely in subjects where my background was limited (most of it) and investigated how various aspects of biology affected the lives of ordinary people. I hoped my approach would engage students and allow them to see that the material they considered as irrelevant while being art, business, or communications majors was indeed relevant to them as consumers, parents, and engaged members of their communities. Student evaluations, the bane of most college faculty, were telling. Some students loved me, a few loathed me, but most grudgingly admitted while they either disliked or had no affinity for science, they learned something. My goal was to be judged by students as tough but fair. A standard I achieved and maintained for three decades.

Invertebrate Zoology was a tough sell. There was little relationship to medicine to attract enough of our majors who leaned heavily to careers in healthcare. I tried different approaches but found students took the class because they needed an organismal course, and Invert was the only class offered at the right time to complete their major. They registered out of necessity, not interest. I loved invertebrates and thought it was important for students to know something about the animals constituting 97% of life on Earth.

I decided to take a different tack. I proposed a course in Tropical Marine Biology, including a Spring Break trip to Jamaica as the field component. Many landlocked colleges offered field courses in marine science to attract students by providing something many kids from the heartland dreamed about but never experienced. I thought it would benefit the college and allow me to continue to introduce students to invertebrates. An approach similar to my wife hiding small pieces of broccoli under the cheese in pizza. The students jumped at the chance to go to Jamaica, but when they got there invertebrates would be the main course. The department was delighted, and the Curriculum Committee approved the class. Invertebrates were out, and Tropical Marine Biology was in.

I knew nothing about marine science when I proposed the course. My excursion to the Gulf Coast Marine Lab was about invertebrates, not the ocean or marine biology. I learned it from the ground up. The old adage is, "you only have to stay one lecture ahead of the students." I read voraciously over the summer. I read the text and purchased almost any book dealing with the ocean, marine plants, or animals. I had little interest in environmental issues before initiating the course, but they were impossible to ignore when dealing with the ocean. Tropical Marine Biology was my introduction to human-based ecological degradation and global warming. Those issues figured prominently in the course I planned to teach. I earned the moniker Dr. Doom. By the time Christmas break concluded, I was ready.

Friends and acquaintances were skeptical of my motives. The consensus was it was a ploy to get a free trip to the Caribbean at the end of winter in South Bend. Not so. I don't like to travel. My attitude is close to heresy but true. I don't care about looking at temples, cathedrals, and castles. I do enjoy the beach and swimming but not enough to get on a plane. I taught the course every other year for 15 years, and invariably this conversation occurred about a week before our departure for the island. Student — "Dr. Platt, are you getting excited about the trip?" Me — "I'll be excited when we get home, you are safe and somebody else's responsibility." The trip to Jamaica was less a perk than a brief excursion to purgatory.

I had to ensure 12–16 young women were safe from the moment we stepped on the bus to go to the airport until our return. We spent a week in a foreign country, subject to their laws, with swarms of good-looking, sexually aggressive young men around every corner. Not to mention the time spent snorkeling, bushwalking, and trying to keep them from doing something stupid, as young adults are wont to do.

I chose the Hofstra University Marine Laboratory (HUML) in Priory, Jamaica, as the site for our field excursion. Eugene Kaplan, from Hofstra

University, founded the lab while compiling a field guide of marine invertebrates in the Caribbean. Gene wanted a place where college students could experience tropical marine ecosystems firsthand. He made a deal with the owner of Columbus Beach Cottages in Priory to convert his facility into a laboratory. The owner was suffering from competition with new luxury hotels going up in Ocho Rios, ten miles down the road, and was eager for the revenue the students represented. Gene ran the science part, and the owner of Columbus Beach provided food and lodging. Gene taught a two-week class for Hofstra students each summer and advertised the facility to other institutions. Faculty from across North America interested in teaching at HUML could take Gene's course to learn about the facility and the area.

I taught TMB six times and took approximately 85 students to the Caribbean without any serious mishaps in large part because Kathy accompanied us as the unofficial chaperone. I doubt she would have permitted me to do the course otherwise. We had some scraped knees and minor illnesses, but nothing remotely approaching catastrophe. The most frightening incident involved a student who had an asthma attack while snorkeling about a quarter of a mile from our boat. Once alerted to the situation, I grabbed the young lady across the chest and towed her to safety. By the time we reached the boat and she was pulled onboard, I was exhausted. I started to sink, and the prospect of drowning flashed through my mind as I dropped further below the surface. I quickly snapped back to reality, released my weight belt, and rose to the surface. It was an eerily odd experience for someone as at home in the water as I am.

We snorkeled in different locations nearly every day, collecting a wide range of invertebrates and algae present in both abundance and diversity. In the evenings, we broke out microscopes and books to identify our catch. The resident directors, hired by Hofstra, gave lectures on marine life, local history, and customs. We took day trips to local markets and other sites of interest.

On each trip, we gathered at dusk in the shallows in front of the lab for the great octopus hunt. We ventured into the inky darkness with headlamps and buckets in search of the reef octopus (*Octopus briareus*). We spread out, searching the nooks and crannies of broken coral. If a student spotted one, she would quietly indicate her find. I moved into position, thrust my arm into the water, hoisted the writhing mass of tentacles over my head, and quickly plopped it into a bucket of seawater. We did our best to keep one pissed-off sea creature contained until everyone saw it. Gingerly its tentacles tested the top of its plastic prison, eventually obtaining a secure purchase and hoisting itself up, out, and back into the water, disappearing in a cloud of ink.

The last two days of the course were hell — on me. In a mere six days, the students collected 100–130 species of plants and animals they had to know for their lab exam. I rolled out of bed between 4 and 5 in the morning and headed to the lab to set up the test — 25 stations with two questions per station. The exam started at 9 am and took an hour. When the exam was complete, the students returned all living animals to as close to the location we collected them as possible and restored the space to its original condition. Whatever remained of the afternoon was free time for the students. I, on the other hand, retired to our lodgings to grade their efforts.

In the evening, we dressed for dinner at a nice restaurant in Ocho Rios or another nearby town. Imagine walking into a crowded restaurant with a dozen, or more, young women dressed for a night out. Heads snapped, and the stares turned to this hoard of lovely coeds. It didn't take long for the wolves to try to separate the sheep from the flock. On more than one occasion, I had to politely discourage potential suitors and shepherd the young ladies back to their tables before anything untoward occurred.

The day of our departure was also fraught. Most of it was mundane, a misplaced ticket or passport, and resolved without incident. Tips were the bane of the course, and HUML ran on tips. Group leaders were expected to tip the staff of Columbus Beach in cash and distribute the money personally. Any local who aided us in any way milled around the compound expectantly. I never questioned the practice. Jamaica is a poor country; local wages low and jobs scarce. These people were kind, attentive, and needed the money. We were more than able to pay for the services rendered. I simply hated the process.

The lab provided a formula for tipping based on group size and length of stay. I came equipped with over a dozen envelopes and lots of cash; ones, fives, and tens. I calculated each person's share, put the money in an envelope, and wrote the individual's name on the outside. I found each person (more likely they found me), thanked them profusely, and handed over the cash. The final bow to local custom was more personal. Our bus driver, Mr. Keize (I never learned his first name), ferried us to all destinations beyond walking distance. Upon exiting the bus at the airport, each student handed Mr. Keize a ten-dollar bill and a kiss on the cheek if so inclined. At long last, we boarded the plane. Back in South Bend, they were safe and somebody else's responsibility. I tolerated it because they loved it.

Parasitology was a different story. Parasitology didn't fill any departmental mandate. It was purely an elective course, meaning students took it because they wanted to, nothing more, nothing less. The discipline divided into medical and zoological factions in the early part of the 20th century. Most colleges and universities started in the zoological camp, teaching the basic biological principles of taxonomy, ecology, evolution, and behavior while downplaying the medical end. I had to design a course meeting the desires of the medically-oriented students Saint Mary's attracted yet retain the basic biological principles I thought important, surprising, and flat-out fun.

I built the class around a slim, medically focused text and included a trade paperback titled *New Guinea Tapeworms and Jewish Grandmothers, Tales of Parasites and People*, authored by Robert Desowitz. The book, published in the early 1980s, provided a series of vignettes of parasitic diseases from around the world with a focus on public health. The most striking theme was the ineptitude of well-intentioned scientists and academics from the first world in attempting to solve the problems of third world communities. An exercise in hubris. We (the smart folks) know more than you (the people living with the problem). Their solutions frequently failed due to the Law of Unintended Consequences. My students loved the stories, and they provided endless opportunities to examine and discuss the hazards of not engaging and listening to those you were trying to help.

The laboratory provided another problem. None of the lab classes I experienced as a student achieved the goals I envisioned. One involved lots of memorization followed by an exam of significant consequence, i.e., a substantial part of the course grade. The other was from my PhD program designed as an independent study geared toward graduate students. I wanted something, while rigorous, that allowed students to experience the wonders of parasitology while keeping the stakes for any single assignment low. The only solution was to write my own lab manual and design exercises students could complete either during the lab or minimal time outside the lab. Each activity carried a small point value, thus eliminating the anxiety of a big lab exam. The result was 11 labs covering approximately 120 pages.

Completing the life cycle of the trematode *Echinostoma caproni* was the focal point. Students had the opportunity to work with living organisms and see all of the stages of the parasite up close and personal. They infected mice and learned how to do fecal exams. They infected snails with miracidia and witnessed the emergence of cercariae, and dissected the snails to find the redial stage. Finally, they examined the mice and found the adult worms in their native habitat — the posterior half of the small intestine.

I designed experiments testing hypotheses of biological significance. They collected and analyzed data and wrote papers explaining what they found and what it meant. I wrote my own keys (identification guides) to the major groups of parasitic worms based on the specimens I had available and provided students unknowns to identify. They examined fecal samples collected from dogs in local animal shelters and prepared blood slides from mice harboring a mouse-specific malaria species. All of the exercises introduced the students to some aspect of parasitological practice or scientific investigation. I stole material with reckless abandon. I followed the old dictum; it is better to seek forgiveness than ask permission. Parasitology was a success. Enrollment was never a problem, and students judged it as one of their favorite courses in end-of-the-semester evaluations.

The decision to focus my research on taxonomy, something I enjoyed and did well, bore fruit. During my probationary period, another six years of uncertainty before a second tenure decision, I published ten papers, all but two on the taxonomy of TBFs[18-20, 23-27]: one of those resulted from a student research project,[21] the second on *Parelaphostrongylus*.[22] I established a coherent research program and a growing reputation in the wider parasitological community. I still didn't have any grants, but Saint Mary's didn't seem to care.

The third leg of the tenure stool, college service, was a harder nut to crack. Most committee assignments open to faculty were the purview of the Faculty Assembly and decided by popular vote. Folks from the human-ities outnumbered the science contingent by a wide margin. I ran for spots every year but always finished out of the running due to block voting by the folks from English, Psychology, Sociology, etc.

I decided to take the bull by the horns and volunteered to serve on the Judiciary Committee — no voting required. Students accused of violating college rules could choose to be judged either by their peers or by us. Why a student would prefer one or the other was a mystery. The cases we adju-dicated were mundane: trying to sneak alcohol into the dorms, trying to sneak young men out, and shoplifting from the bookstore come to mind.

The screen slasher incident was the most memorable. One evening three students returned to campus after a night of partying. It was very late (or early), and upon opening the door, the students turned on the lights waking their two sleeping roommates. Their room was in an old building, constructed in the early 1900s, with 12-foot ceilings and occasionally vis-ited by wayward bats. That evening there was a bat in the room. Screaming ensued, and someone called Campus Security to deal with the problem.

An officer arrived and attempted to swat the bat with a broom, a highly ineffective approach. While four of the five young ladies huddled under their bedcovers and the officer played *Casey at the Bat*, one intrepid student took matters into her own hands. She heard if the bat sensed a way out, it would leave. She went to a window and tried to remove the screen providing the bat egress. Unfortunately, the screen, firmly affixed to the window frame, wasn't coming off in the absence of a screwdriver. With that avenue closed, the resourceful undergraduate grabbed a pair of scis-sors and slashed an 'X' in the screen, pushed it open, and the bat departed. The Resident Assistant wrote her up for destruction of college property.

Our panel listened intently as both sides shared their stories. We thanked them and requested they step into the hall while we deliberated. As soon as the door clicked shut, we all broke out in peals of laughter, which I was sure were audible to the litigants waiting outside. We called them back into the hearing room, dismissed the charges, and apologized to the accused for everything she endured. We also assured everyone pre-sent the action taken was the best solution to the problem, and the college could bear the cost of replacing the screen.

Our department consisted of eight faculty, each with his or her specific area of expertise. Most of us were of similar age (30–40) but with wildly different religious views, political proclivities, and personal interests. We worked marvelously together and went our separate ways at the end of the day. During my nearly 30 years, there were no turf wars, power struggles, or personality conflicts. We did have disagreements, but we were always able to discuss matters calmly and arrive at a consensus. The overarching question determining any decision was, "What is in the best interest of the students?" With student welfare as our North Star, personal wants and needs were incidental to providing the best possible educational experience for the young women in our classes. It was a terrific place to work.

The years flew by, and I prepared for my tenure application in 1991. The process at Saint Mary's differed slightly from what I experienced at Richmond. At Richmond, the department voted and sent their recommendation to the Dean. At Saint Mary's, each department member wrote an individual evaluation to the Committee on Rank and Tenure, composed of three faculty members elected by the Faculty Assembly, the Dean, and a Presidential appointee. They evaluated the material and made a recommendation to the President, then from the President to the Board of Trustees. I devoted the summer of 1991 to compiling my portfolio: accomplishments in teaching, research, and service. I solicited letters of recommendation from colleagues at Saint Mary's to assess my service and parasitologists at other institutions to evaluate my research. All materials were due in the Dean's office by October 1st, and notification of the decision would arrive by mail on Valentine's Day.

I felt confident, but I felt confident at Richmond too. If I was nervous, Kathy was a wreck. She hid it well. I had no idea the depth of her anxiety until I arrived home from work on February 14th, 1992. Mail typically arrived between 1–2 pm. I got home around 6. Kathy handed me the letter, and stupidly I asked, "You didn't open it?" Of course, she hadn't opened it. I slit the envelop and read as far as, "I am pleased…" Kathy started crying. We hugged and kissed. She told me she had been holding her breath for six years. Now she felt we had a home, a place where we belonged for the long term, and she could quit worrying about the future. Valentine's Day remained a day to express my love for my incredibly supportive wife and not tarnished by another rejection by the academy.

I was eligible for a sabbatical the following year and began to fantasize about places to go and what I needed to do to get there. My recent contact with scientists 'Down Under' (the publication of the description of *Griphobilharzia amoena*[25] — see Chapter 11) gave me the destination — Australia. If I wanted to make the dream real, I needed to break my long-standing aversion to grants. Most applicants don't receive funding on their first attempt, but I didn't have that luxury. I needed to score on the first try. I needed a rationale for going and a funding source with a high probability of success.

The Lilly Endowment, located in Indianapolis, was my best bet. The foundation supported a wide range of programs, but my target was the Lilly Open Fellowships. Each year they offered ten grants of up to $40,000 to college faculty from Indiana and only Indiana. The money, while not eye-popping, was sufficient, and the competition local. The program's principal objective was the proposed project provide some benefit to the home institution when the grant ended. I needed a reason for them to give me the money.

I decided learning the fundamentals of molecular biology and its application to taxonomic studies was my best bet. I established myself as a productive taxonomist using traditional morphological approaches. The addition of molecular techniques would enhance my research program. Most importantly, Saint Mary's did not offer molecular biology, the fastest growing and most powerful addition to the biologist's toolbox. David Blair, my co-author on the *Griphobilharzia* paper, was an expert using molecular biology in trematode systematics, so I needed him to agree to teach me the basics. He was open to me working in his lab, so the circle was complete. I had the place, James Cook University, Townsville, Australia, and the project, molecular assessment of turtle blood fluke (TBF) taxonomy. All I needed was the Lilly Endowment to fork over the money.

After several months of writing and polishing, I dropped the finished proposal in the mail. I waited to learn if I was a finalist. Finalists appeared, in person, before a panel at the endowment headquarters in Indianapolis for a Q & A. I made the cut, and the date and time for my interview was scheduled for 10 am in mid-November. The drive from South Bend to the Lilly Endowment headquarters was a hair over two hours. I was out the door at 5 am. Nothing like a bit of a cushion for possible car trouble, jack-knifed trucks, or an act of God. The drive was closer to 2 ½ hours, and I spent the bonus time sitting at an IHOP two blocks south of the

Endowment headquarters reviewing my proposal and rehearsing answers to questions I thought they might ask.

The interview room was beautiful, old-money wood paneling with the requisite portraits of departed members of the Lilly family, a large conference table, and overstuffed chairs. I was at one end of the table and my interrogators at the other. After a few minutes, I began to relax. The questions were softballs, and the folks at the other end of the table were smiling and nodding their heads in approval at everything I said. I couldn't say the fix was in because I didn't know any of the Lilly crew. It seemed they wanted to give me the almost $40,000 I requested. It was all smiles and handshakes when I left. They still had more interviews, but I should hear within a week, they assured me as I headed for the parking lot and home. The letter arrived ten days later. I got the grant, and our family was going to Australia.

July was nearly over, and our August departure date fast approaching. I shipped half a dozen boxes of books and supplies earlier in the summer in anticipation of our arrival, so we only needed to deal with clothing and other personal items to carry on the plane. The problem was our children. Neither was happy and certainly didn't see spending a year in Australia as the adventure of a lifetime. It didn't matter. We drove to Chicago, handed our car to friends, boarded the plane for Los Angeles, and on to Hawaii. We were allowed one layover in each direction, and Kathy thought this was an excellent opportunity for the kids to see our 50th state. The three-day stay in Honolulu broke up the 20-hour flight, with Australia on the horizon.

A couple of weeks before our departure, a short article appeared in the *South Bend Tribune,* stating the president of Wabash College, in Crawfordsville, IN, a 2 ½ hour drive southwest of South Bend, resigned with no immediate plans. A polite way of saying he was fired but allowed to bow out gracefully. The president's name? Sheldon Wettack. The same Sheldon Wettack who fired me at Richmond. I had a brief moment of schadenfreude, but I let go of my anger at him years earlier. He did me a favor. We loved South Bend, I loved Saint Mary's, and my career was on an upward trajectory.

10. Turtle Blood Flukes

This is where it all begins. Everything starts here, today.

— David Nicholls

Why the Spirorchidae, or turtle blood flukes (TBFs)? First, they are cool as hell. The first time I saw *Spirorchis scripta* undulating in blood vessels of a painted turtle I dissected for my Master's thesis, I was enthralled. These worms are common in pond turtles in the Midwestern United States, and I had the opportunity to observe them during the months of collecting specimens at BGSU. The lakes and ponds near South Bend offered an abundant supply of turtles, and where there are turtles, there should be TBFs. While I started locally, TBFs would take me around the world.

Several factors steered my decision. First, the group had been neglected for decades. Horace Stunkard was the last person to study them intensely, and his work spanned 1921 to 1928. A few folks dabbled with them during the ensuing years, but no taxonomist claimed them as their own. I had an open field. Second, almost everything published on this group was in English. I took Spanish in high school and learned to read Russian in graduate school, but both faded quickly from disuse. Finally, I figured there were lots of new taxa awaiting discovery because most turtle species had never been examined for these fascinating creatures.

My decision proved to be career-defining. 25% of my publications in the next 30 years dealt directly with TBFs and provided an entrée to

Australia, which resulted in an additional ten non-TBF papers. My decision to focus on this neglected group of trematodes directly or indirectly resulted in 40% of my publications.

I had hoped to get a paper out quickly after arriving at Saint Mary's. The editor of the *Journal of Parasitology* rejected my first attempt at TBF taxonomy. Still, I strongly felt the material I found in snapping turtles at the University of Richmond was new. I retrieved the manuscript from my files, read it carefully, and decided what improvements were necessary to get it accepted. It was clear I hadn't made my case. I needed to obtain comparative material to strengthen my argument. I contacted the directors of parasite museums worldwide and requested specimens deposited by previous workers. And, yes, there are parasite museums.

The primary repository was the United States National Parasite Collection (USNPC, now part of the Smithsonian) in Beltsville, Maryland, headed by J. Ralph Lichtenfels, who served as the external examiner for my PhD defense. Other specimens were in collections across the globe: England, Japan, Malaysia, and, unfortunately, India. Indian parasitologists, for reasons I can't fathom, kept their specimens in personal collections and steadfastly refused to allow me, or anyone else, to borrow them for study. I complained about this practice in private correspondence and in print to no avail.

Museum curators responded positively to my requests, and specimens arrived in my mailbox. The Indians ignored me or begged off with myriad excuses for why my request was impossible to grant. The most poignant letter I received was from the Hiroshima Museum of Natural History, informing me that during World War II, the detonation of the atomic bomb vaporized the material I requested.

The only significant taxonomic work on TBFs between Horace Stunkard and my decision to focus on them appeared in 1939 by Elon E. Byrd from the University of Georgia. Byrd studied turtles from Reelfoot Lake in Tennessee. He described several new species and constructed a crude evolutionary scheme for all of the blood flukes. Byrd was a consummate professional and deposited representatives of his finds at the USNPC; however, much of the material was in poor condition and not as helpful as I hoped. I had no doubt there were more specimens in his private collection. The only problem was Byrd died in the early 1970s. Inquiries to his friends and colleagues failed to shed light on the remainder of his collection.

Eventually, I learned Byrd's specimens never left the university. My hunch was correct. There were numerous additional specimens from Byrd's Reelfoot Lake study, and to my surprise, many others of taxonomic significance. I received permission to remove them on the condition that I write a paper indicating what I found and take the step Byrd should have — deposit them in accredited museums.[23] In all, I rescued over 350 slides representing more than 30 species from a storage area in the University of Georgia vet school. I have no idea what became of the remainder. Curating the thousands of slides I left behind would take years, time I didn't have to devote to such an undertaking. I hope somebody found them a proper home.

I resurrected the rejected manuscript describing a new species of the genus *Hapalorhynchus* and got to work. I compared it to the original specimens of the type species, *H. gracilis*, described from snapping turtles from North Judson, Indiana, less than an hour's drive from South Bend. I examined a species, *H. stunkardi*, described by Elon Byrd in his treatise from Reelfoot Lake. I prepared specimens of the new species for sectioning to elucidate the nature of the reproductive system for a genus-level review. While the paper's primary objective was to describe a new species, *Hapalorhynchus brooksi* (Figure 2), named in honor of my friend and colleague, Dan Brooks, I also provided redescriptions of *H. gracilis* and *H. stunkardi*. This more thorough, comparative approach to the problem was viewed favorably and published in 1988.[18]

So what is a species? It would be nice to tell you there is a single, agreed-upon definition of this foundational biological concept, but there isn't. The last time I looked, there were about 30 of them. For the first 100 years of post-Darwinian taxonomy (and before), the foundation of species determination rested on morphology. Ernst Mayr introduced the Biological Species Concept in the late 1950s. Mayr defined a species as an interbreeding population of organisms that don't interbreed with members of any other population. Sex was the defining character. The drawback was it eliminated fossil and asexual organisms and most of the worms we study. Trying to determine reproductive compatibility between worms collected in Indiana with similar worms collected in Virginia is

Figure 2. *Hapalorhynchus brooksi* from: Platt, T.R. 1988. *Hapalorhynchus brooksi* sp. n. (Trematoda: Spirorchiidae) from the snapping turtle (*Chelydra serpentina*), with notes on *H. gracilis* and *H. stunkardi*. *Proceedings of the Helminthological Society of Washington* 55: 317–323. (Reprinted with permission.)

impossible. There are other problems with this approach, but we will leave them for the philosophers and cognoscenti.

DNA promises to be the great leveler — maybe. Differences in nucleotide sequences and SNPs (single nucleotide polymorphisms) can be enumerated and compared across time and distance. How many nucleotides or SNPs have to differ to distinguish one thing from something else? And if all you have are DNA samples, they may be different, but how do you attach them to an organism? Those issues remain unresolved. We may get an answer at some point, but we aren't there yet.

My favorite definition is "a species is whatever a competent taxonomist says it is." Two competent taxonomists might disagree how different the morphological differences need to be or how much nucleotide variability is necessary to say two things are different. Until a gold standard is found and agreed upon, I see no clear resolution to this conundrum other than the subjective assessment of people who devote their lives to a group of organisms.

During the next five years, TBFs formed the focus of my personal research. I employed cladistics to assess the evolutionary relationships of the species of *Hapalorhychus*[18,19] (typically found in snapping turtles and musk turtles) and *Spirorchis*[27] (parasites of various pond turtles). I also reviewed and reassessed the morphology of all of the recognized species of *Spirorchis* from North America.[29] I was making a name for myself in parasitological circles.

Finding new taxa does not require trips to exotic locales; sometimes, they are right around the corner. The old farmhouse Kathy and I purchased on the south side of South Bend was a quarter of a mile from a small, unnamed pond at the corner of Lilac and Johnson Roads. It was highly eutrophic (lots of plants and decaying vegetation on the bottom) and was home to a cornucopia of turtles and snails and, I hoped, parasites. I regularly trekked to the pond in chest waders, carrying a dip net, pillowcases, and plastic bottles to transport my catch back to the house. Passing motorists could only guess what I might be doing in that outfit. The most common turtle species were snapping turtles and the smaller, less ferocious painted turtles. Both harbored myriad worms in the intestine, lungs, urinary bladder, blood, and other organs. Most were species with which I was intimately acquainted.

One afternoon, I caught a small painted turtle in my dip net while sampling for snails. It was about the size of a half-dollar. I decided to keep it in a small aquarium on my desk with no intention of examining it. It did everything you would expect of a turtle. It ate the commercial turtle food I offered and basked on a small island I constructed in the tank. The little bugger dove for the safety of the bottom when I waved my hand over it while it was sunning, only to resume basking when the apparent danger passed. After a few weeks, I decided to see what parasites this tiny turtle might harbor.

The necropsy proceeded along typical lines. Once the animal was dead, I removed the plastron (the bottom part of the shell) and separated the organs into individual Petri dishes for examination. There wasn't much, and I felt a pang of remorse for killing that young animal. However, something caught my eye. Instead of their normal light pink color, the lungs appeared golden brown under the light of the dissecting microscope. I cut a small section of the organ and placed it on a microscope slide for closer examination. The color was due to massive numbers of parasite eggs clogging the pulmonary blood vessels. The turtle must harbor TBFs, but there was no trace of the adult worms. Eggs also peppered the lining of the intestinal tract. With this many TBF eggs, where were the parents?

I opened the skull and found the answer. The blood vessels on the surface of the brain were lined with eggs, and the telltale sign of adult worms was clearly evident; black pigment characteristic of digested blood in the intestinal cecae (the gut) of the worm. Removing the adults from the blood vessels was a nightmare. They were long, thin, and tightly wedged in their tubular abode. I eventually removed half a dozen intact adults from the brain of my tiny turtle. I found additional specimens in other painted turtles during the next month.

Comparison of my new find with the previously described genera and species of TBFs confirmed my suspicion — it was a new genus and species.[24] And I found it in a pond a short walk from our house! Preparation of a manuscript proceeded quickly. I christened this new worm *Aphanospirorchis kirki* (Figure 3). *Aphano-* is Greek for unseen, and *spirorchis* denotes a TBF — the unseen spirorchid. The specific name, *kirki*, honored Daniel Kirk, a kindred spirit who did his PhD work on these gorgeous animals and donated his specimens to me for study.

During my stay in Australia (Chapter 13), David Blair gave me access to his collection of TBFs from marine turtles. David studied trematodes in the family Pronocephalidae (among others), common inhabitants of the intestinal tract of turtles, and many other vertebrates. David amassed an extensive collection of TBFs while dissecting turtles found dead, or dying

Figure 3. *Aphanospirorchis kirki* from: Platt, T.R. 1990. *Aphanospirorchis kirki* n. gen., n. sp., (Digenea: Spirorchidae) a parasite of the painted turtle, *Chrysemys picta marginata*, from northwestern Indiana, with comments on the proper spelling of the family name. *Journal of Parasitology* 76: 650–652. (Reprinted with permission.)

on the beach, or that had been mortally wounded by natural predators, or encounters with boat propellers. David was working on other projects and was delighted to hand off the TBFs to me.

Some of the specimens were already stained and on slides; others were still in vials and required processing for study. I did some preliminary work on this collection in Australia, packed them up, and shipped them back to Saint Mary's for a more thorough examination.

Those specimens formed the basis of two publications. One cleared up a technical issue boring to all but a small cadre of biologists who attempt to understand the highly legalistic International Code of Zoological Nomenclature.[33] In essence, I was arguing which name most appropriately belonged to the type species of the genus *Hapalotrema*. Pretty exciting stuff, eh? I tackled two nomenclatural issues in my career and came away with a splitting headache both times. I embarked on the second paper, reassessing the species of the self-same *Hapalotrema* and clarifying a series of confused and incomplete descriptions by earlier workers.[35] I should note as technology improves and standards rise, work deemed acceptable decades earlier is no longer adequate. Science progresses in how we study and understand problems. Revision is inevitable. I don't doubt that some things I published will be deemed confused and incomplete by a younger version of myself at some point in the future.

The last paper I wrote before embarking on the Australian material I collected on sabbatical (Chapter 12) was a thought experiment.[34] Most trematodes are monoecious, i.e., hermaphrodites. Each individual contains functional male and female reproductive systems. Hence they inseminate other individuals and are themselves inseminated. This is a common phenomenon in animals of low vagility (i.e., they don't move very far, very fast) or are rare. In either case, the chance of meeting another individual of the opposite sex is low. A male encountering another male (or a female another female) after an exhaustive search for a mate would mean evolutionary death: no offspring, no genes passed to the next generation — a wasted life. But if a hermaphrodite meets another hermaphrodite, they can swap sperm and be on their way. Even if you don't find someone to canoodle with, you might be able to fertilize your own eggs. Not an ideal situation, but better than having no offspring at all.

Before the turn of the new millennium, three families of blood flukes comprised the superfamily Schistosomatoidea (all of the blood flukes of vertebrates): the Aporocotylidae, parasites of fish; the Spirorchidae, parasites of turtles; and the Schistosomatidae, parasites of birds and mammals (and crocodiles!). The first two families are hermaphrodites; each individual possesses both male and female reproductive organs. The third, the schistosomes, are dioecious with separate sexes. Dioecious trematodes are rare. How and why the condition evolved from monoecious ancestors was (and still is) a burning question in parasitology. I thought I would take a crack at it. Numerous authors published papers speculating on the subject, but there is no definitive answer. I didn't think I would solve the problem, but I might nudge the ball forward.

To appreciate my thinking, it is necessary to understand how blood fluke eggs escape the host for each of these groups — they are similar but differ significantly. The adult worms live in the heart and blood vessels of their respective hosts. Escape is neither direct (as is the case for worm eggs in the intestine leaving via the poop) nor easy.

The fish heart consists of two chambers, the atrium (which receives blood from the body) and the ventricle (which pumps blood to the body). In fish, the blood in both chambers is low in oxygen. The atrium passes blood to the ventricle, which sends it to the gills to release carbon dioxide and pick up oxygen. The small vessels in the gills (capillaries) are narrow, forming a trap for any worm eggs released by the adults living downstream. The eggs of aporocotylids either work their way through the capillary and gill tissue and develop in the external environment or produce a miracidium in the capillary, which escapes in search of an intermediate host.

Turtles, birds, and mammals have a four-chambered heart. The arterial and venous systems are separate. The right side transports deoxygenated blood from the body to the heart and from the heart to the lungs. The left side transports oxygenated blood from the lungs to the heart and from the heart to the body. The position of the adult worm is critical for the eggs to leave the host as the eggs lodge in the first capillary bed they enter. TBFs tend to reside in the arterial system of their turtle hosts. Their strategy is similar to plants using wind to disperse their pollen. The wind blows and the pollen lands whereever the air currents carry them. The same is true of TBF eggs. Since oxygenated blood is sent to all body tissues by arterial blood flow, TBF eggs occupy nearly every organ in the body. Eggs trapped in the capillaries of the gut are most likely to make their way from the blood vessels into the intestine and exit in the poop! The rest eventually perish. The exact mechanism of the movement of the eggs from the circulatory system to the lumen of the gut, while critical to the worm, is not something we need to consider now. The massive number of eggs I encountered in some turtles during my career suggests killing a turtle is not easy. The little turtle infected with *Aphanospirorchis kirki* behaved normally despite having its lungs and brain outlined with TBF eggs.

The schistosomes found in birds and mammals are dioecious, and most demonstrate sexual dimorphism. The males are larger than the females. Early in their lives, they form pairs lasting (as far as we can tell) a lifetime, which may be a decade — or more. They have to find a partner to develop to sexual maturity. So while they are dioecious, they are functionally hermaphrodites.

Schistosomes employ a radically different strategy. Schistosomes reside in the venous, not the arterial system. Letting their eggs disperse at the whim of blood flow is not an option. The females place their eggs in locations with a high probability of exiting the host. Living in the veins may seem counterintuitive. Oxygen levels are low, and blood flows away from the organs where the eggs might escape the host and back toward the liver and heart. An egg swept back to the liver would have little expectation of ever escaping the host. This is where sexual dimorphism is critical.

The males are large and muscular, while the females are long and thin. The male carries the female down a vein as far as his size allows, then the female leaves the male, like a shuttle leaving the mother ship. She moves into smaller and smaller vessels before depositing her precious cargo. The eggs are slightly larger than the vessel, and the elasticity of the venule (= small vein) holds the eggs in place. The female carefully backs out, laying eggs as she returns to her partner. Precise placement of eggs in the small veins of the intestine or urinary bladder (and other locations) ensures at least some eggs are held firmly in place by the vessel and have a reasonable chance of escape.

Once the process is complete, and the couple reunites, the male carries the female to a new vessel, and the process repeats. The eggs release molecules stimulating an inflammatory response. The eggs move slowly toward the lumen of the gut or the bladder and are expelled like a splinter works its way out of your finger if left unattended. The other half? Their purchase in the vessel is less secure, and they flow back into the liver. Eggs in the liver are a problem, but not an immediate one. The liver can absorb a great deal of damage before it fails and the host dies. In the meantime, thousands of eggs leave the host. It may be years or decades before the host succumbs to a failing liver.

Another advantage of living in the venous system is the increased availability of nutrients, especially if the worms live in the veins transporting blood from the intestine. This portion of the circulatory system is called the hepatic portal system. It carries blood from capillaries in the gut to capillaries in the liver for detoxification before the blood reaches more sensitive parts of the body, and is extremely rich in nutrients. This nutrient-rich environment provides the energy for the high fecundity these worms exhibit.

During my stay in Australia, I found two distinct forms of the genus *Uterotrema*, a TBF, in the Australian snapping turtle, *Emydura krefftii*.[32] One was small, thin, with a large uterus and numerous eggs. The other was larger, more robust, with a small uterus devoid of eggs. While I described them as separate species (*U. burnsi* and *U. krefftii*, respectively), I hypothesized (i.e., guessed) they could be morphs of a single species. Maybe the worms begin life as small, thin hermaphrodites, but as they grow and age, they become functional males preferentially inseminating the smaller hermaphrodites. Eventually, the genes for maleness cease functioning in the hermaphrodites. The genes for femaleness do the same in the larger protomales giving rise to the dioecious condition found in the schistosomes. Recent molecular genetic studies suggest this may be the way monoecious blood flukes became dioecious. The jury is still out. Only time will tell if my guess will bear fruit or end up in the dustbin of crackpot ideas littering the scientific literature.

I was incredibly proud to be invited to write the chapter on TBFs for a multi-volume key to the genera of trematodes edited by David Gibson and two other colleagues from the Natural History Museum (London).[43] I had to review the literature on all the genera and species, assess the characteristics allowing unequivocal identification of each genus from every other genus, and construct a set of paired choices to accomplish the task. I would also have to prepare original drawings of the type species for all 20 genera comprising the family. To be fair, other friends and acquaintances authored multiple chapters, most containing many more genera than the small family of TBFs. Our reward for this effort? A free copy of the volume to which we contributed valued at about $250. It was the only time I received compensation, however small, for anything I published. It was, for me, a significant undertaking and cemented my status as the expert on this small group of trematodes.

I wrote four more papers on TBFs before I retired. One was a redescription of a species described by Elon Byrd back in 1939,[49] requiring an update. Two were from my time in Malaysia and reviewed in Chapter 22 and *Baracktrema* (Chapter 1). The final paper was a foray into the early history of Parasitology in the United States.[70] An attempt to understand the relationships between four strong-willed individuals based on a series of 50 letters bouncing between them in the early years of the 20th century. When I began the project, I gave no thought to publishing the results. I wanted to understand their interactions and motives. It was an itch I needed to scratch. The deeper I got into these people's lives, the more I thought other folks might be interested as well.

11. The Worm Where Nothing is Where It's Supposed To Be[25]

The real voyage of discovery consists not in seeking new landscapes, but in having new eyes.

— Marcel Proust

Where were you when you learned of the assassination of President Kennedy? When the Twin Towers fell? Iconic moments are indelibly stamped geographically for reasons not entirely clear. Other, more personal events that lack the gravity of national tragedies may also resonate.

I attended the Fifth International Congress of Parasitology (ICOPA-V) in Toronto, Canada, in 1982. The city changed dramatically since my last visit a decade earlier. Elected officials undertook a massive cleanup campaign during the previous decade, resulting in a brighter and cleaner look for the city center's franchise hotels.

International meetings are huge, drawing thousands of scientists from across the globe. Attendees at ICOPA-V gathered at the Sheraton, a massive and opulent structure. Scientific meetings are about the formal transmission of discoveries in oral and poster presentations and informal exchanges in private conversations in hallways, restaurants, and bars. As a newly minted PhD in my early 30s, my cohort spent more than our fair share of time knocking back beer and sharing war stories about our graduate programs, gossiping about our elders, and talking shop.

One afternoon, during a break in the presentation schedule, I was in the Sheraton bar with three friends: Dan Brooks, Bill Font, and Tom Deardorff. Bill is quiet, unassuming, and uber-smart. His demeanor belies the fact that he was a fighter pilot in Vietnam. Tom, a Magnum PI look-a-like, was employed by the Food and Drug Administration and chaffed against the strictures of working for the government. Dan was the unquestioned leader of the group: a significant presence, both physically and scientifically. Dan attended the University of Nebraska on a track scholarship and published 30 papers by the time he finished his PhD. Dan introduced cladistics and vicariance biogeography, new ways of determining evolutionary relationships between living organisms and understanding their distribution on the planet, to our discipline. He had no qualms about taking on the old guard and challenging traditional approaches to parasitology.

During our conversation and several rounds of beer, Dan mentioned he had recently visited David Blair at Otago University, New Zealand. David showed him a parasite found in freshwater crocodiles from Australia's Northern Territories that baffled them both. Dan is an excellent taxonomist and extremely knowledgeable about trematode morphology. Most of Dan's publications included descriptions of new trematodes and cestodes. I listened to his story of this baffling organism with rapt attention. I recall Dan saying, "Nothing is where it's supposed to be." If Dan couldn't figure this thing out, it must be bizarre. I didn't know anything about David, but our lives would intertwine in the near future.

Fast forward six years. I had been fired from my position at the University of Richmond, worked for Midas Rex for six months, and was two years into my new job at Saint Mary's. I was getting settled into the culture of a new school, new colleagues, and a radically different student clientele — all women. I also made a change in my research program; the taxonomy of TBFs.

Out of the blue, a letter arrived from David Blair, James Cook University, Townsville, Australia. I didn't know anyone in Australia personally, but the name, David Blair, sounded familiar. I loved getting mail at work. Letters at work were fun. Bills arrived at home, but the mail at school was about science. I opened the letter and read. David had a strange worm veterinarians in the Northern Territories collected from freshwater crocodiles. It was from the circulatory system and was, therefore, most likely a TBF. He heard I was working on those critters' taxonomy, and he had neither the time nor inclination to pursue this strange beast further. Was I interested?

David described the basic morphology as he saw it and included a handmade drawing labeling what he thought the various structures might be. There were notations, mostly question marks. As I read the description and examined the drawing, I realized this was the worm "where nothing is where it is supposed to be" Dan described over beer six years earlier. I wasn't sure what to think. I certainly didn't have the stature to engender the trust David offered to place in my hands. If neither Dan nor David could crack this enigma, what chance did I have? Curiosity won out over self-doubt, and I wrote back telling David to send the specimens. I would do my best.

Two months later, a small brown package arrived from Down Under. I took it back to my office and briefly stared at it before opening my pocket knife and carefully slitting the brown butcher paper. The package contained a small slide box, a bit bigger than a deck of cards with slots holding a dozen microscope slides. Each slide was wrapped carefully in a single sheet of paper cut to size. The spaces between the slides were stuffed with cotton to protect them from moving and breaking during their transit through the vagaries of the Australian and U.S. postal systems. There were also two dozen or so small vials containing preserved specimens. I needed to process those personally and make my own slides. David included records documenting the host (size and sex), locality (where the croc was captured), date of collection, and the location of the worms in the host at necropsy. He also added a few observations; most notably, the worms were hard to stain.

Staining is a process of exposing a specimen to different dyes to enhance the worm's internal anatomy. Different organs take up different stains at different concentrations. Staining is akin to using invisible ink. The untreated paper looks blank, nothing of interest. However, when you add lemon juice and a bit of heat, the hidden writing comes clear and is easily differentiated from the surrounding white. Some species of trematode stain easily and make beautiful specimens; others — not so much. Why? I have no idea; they do, or they don't. You find out by trial and error.

There are myriad different stains, primarily developed in the late 19th and early 20th centuries. Some work well on some species and not on others. Why? Again, I have no idea; they do, or they don't. Staining is more art than science.

The worms were small, about one-tenth-inch long, the width of a human hair, and extremely delicate. During staining, specimens are transferred from solution to solution using either a fine-tipped artist's paintbrush or a pipette. Both pose hazards. The specimen can become entangled in the brush or stuck inside a pipette. I damaged or lost worms to each. I finally managed to obtain a half dozen mounted specimens for microscopic examination.

Having cleared the technical hurdle, I tried to decipher the worm's anatomy and where it belonged in the trematode hierarchy. It was undoubtedly new, but a new species? A new genus? A new family? I placed a slide on the microscope stage and positioned the objective lens carefully to avoid damaging the specimen. As I increased the magnification from 40× to 100×, I was able to identify familiar landmarks using David's rough drawing and my knowledge of trematode anatomy. The oral sucker and ventral sucker were clearly visible. I could make out the esophagus and a pair of short intestinal cecae — the worm's digestive system. At the posterior end, there was a structure that appeared to be a single testis, but the middle of the worm was a mess. There seemed to be two large spaces and structures I couldn't relate to the standard trematode model. In several specimens, a mass of eggs clogged the gap anterior to the thing I thought was the testis. I examined each worm and ended up in the same place. The easy stuff was easy, and the rest made no sense. Nothing changed during the following year. When I finished preparing lectures, advising students, and working on projects moving more rapidly toward completion, I went to the lab to examine the specimens for an hour or so until my back or eyes gave out. I was close to telling David I was done and would return his specimens.

Isaac Asimov, noted author, commented on the scientific enterprise with a quote I regularly shared with my students: "The most exciting phrase to hear in science, the one that heralds discoveries, is not 'Eureka!' (I found it!), but "That's funny…." One quiet afternoon, like Sisyphus, I returned to my microscope to attempt to roll my wormy stone up the hill.

I moved quickly to a magnification of 400× and saw nothing I hadn't seen before. I moved to 1000×, the highest magnification my scope could muster, using oil immersion. A drop of oil between the condenser (light source) and the bottom of the slide, and between the specimen and the objective lens on the top, helps focus the light and improves the sharpness of the image.

I used oil in the past, so I wasn't optimistic about any outcome other than the utter frustration experienced on so many prior occasions. I looked at an area around the "gap" near the testis (a fact I confirmed by detecting sperm in the organ). I noticed something I hadn't seen before. And then I had my "isn't that funny/Eureka" moment. I realized I was looking at a sucker, a sucker in a place where a sucker shouldn't be. Further examination revealed a tube through the sucker connected to an esophagus, and I knew what I was seeing: a second worm. A worm inside a worm! The outline of two faces magically became a vase.

The blood flukes of fish and turtles are hermaphrodites (monoecious). The schistosomes found in birds and mammals have separate males and females (dioecious). The "crocodile" worm had to be a schistosome because they were dioecious. There were males and females, but it wasn't like any other schistosome because the two worms were oriented "head-to-toe" rather than "head-to-head" like all other described species. Rather than lying free in the male's ventral groove, the female of this bizarre creature resides in a chamber in the body of the male — a relationship strange beyond belief! I began a careful examination of all the specimens, identifying the male and female organs in each pair of worms and measuring them in preparation for formal description and publication.

Further examination of histological sections across the body of one specimen confirmed the male surrounded the female. The enclosed female had no means of escaping her suitor. There was no sign the male's lateral margins wrapped around her and fused, and even the smallest specimens, far from sexual maturity, showed the female developing inside a male. Questions about this bizarre situation arose almost immediately: Do the two sexes start as independent individuals and come together early in development, or does a single egg contain the developmental instructions for separate sexes? How does the female obtain nutrients? How do the eggs, which appear to collect in the male chamber, escape from the worm, and make their way to the outside to continue the life cycle? I had no answer for any of them.

Naming this new find was my most ambitious venture into the nomenclatural realm. I searched guides to classical language, looking for terms reflecting the unique nature of this trematode. After several weeks, I settled on *Gripho-* which translates as "a riddle" and *bilharzia* in tribute to Theodor Bilharz, a German physician who first reported schistosome eggs in the bodies of cadavers during a visit to Egypt in the 1830s. This suffix is commonly used in medicine to describe the disease caused by schistosomes — bilharziasis. The species name *amoena* translates as charming. I christened our discovery *Griphobilharzia amoena* (a charming schistosome riddle — Figure 4). A most fitting sobriquet.

The manuscript was accepted for publication,[25] and following accepted practice, I deposited the specimens in several different museums. I sent the holotype (the specimen illustrated in the article) to the South Australian Museum, Adelaide, Australia. Additional specimens went to the United States National Parasite Collection, Beltsville, Maryland (USNPC, now located at the Smithsonian), and the Harold W. Manter Museum, University of Nebraska-Lincoln. Having specimens in multiple locations made them available to a wide variety of researchers, and if tragedy were to strike one institution, all was not lost.

The early '90s marked the beginning of the molecular revolution in taxonomy, systematics, and evolution. Morphology was the benchmark for these disciplines since Darwin's publication of *The Origin of Species* in 1859. With the discovery of the structure of DNA by Watson and Crick in 1953, the molecular revolution was born. Comparison of DNA sequences is now a staple in identifying new taxa, assessing their genealogy, and tracking evolutionary history. As the decade rolled on, I received regular requests for specimens for DNA analysis. As the first "schistosome" found in a reptile, *Griphobilharzia* appeared to hold answers linking the TBFs (Spirorchidae) to their relatives found in birds and mammals (Schistosomatidae), including species causing the debilitating disease schistosomiasis in humans. Unfortunately, all the original specimens had been fixed in formalin, which rendered them, at the time, useless for molecular analysis. If a DNA comparison of *Griphobilharzia* was going to happen, someone would have to trek to Australia's Northern Territory and collect new material.

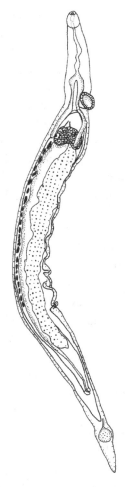

Figure 4.   Griphobilharzia amoena from: Platt, T.R., D. Blair, J. Purdie and L. Melville. 1991. *Griphobilharzia amoena* n. gen., n. sp. (Digenea: Schistosomatidae), a parasite of the freshwater crocodile, *Crocodylus johnstoni* (Reptilia; Crocodylidae) from Australia, with the erection of a new subfamily, Griphobilharziinae. *Journal of Parasitology* 77: 65–68. (Reprinted with permission.)

In a cruel twist of fate, the original specimens came from "farmed" crocs raised for their hides and meat. The worms caused problems for the farm owners, and they filled in any ponds containing the parasite and exterminated the infection. Attempts to control parasitic infections are notoriously ineffective, but not in this case. The croc farmers drove

Griphobilharzia amoena to extinction in its only known locality. Finding new sources of material would require catching infected crocs in the wild.

A little more than a decade later, a group from the University of New Mexico headed by Eric Loker traveled to Australia and collected new specimens of *Griphobilharzia*. Their molecular analysis was nothing short of stunning. Rather than showing the croc worm closely related to the dioecious schistosomes where David and I placed it in our original publication, it nested right in the middle of the monoecious TBFs; not the first time molecules and morphology disagreed on the evolutionary placement of an organism. Unfortunately, I do not have the expertise to appreciate fully or challenge the molecular data and the University of New Mexico group's incongruous result.

Occam's razor is the only tool available to me, which states the simplest answer is usually the best. In this case, Occam would support the placement of *Griphobilharzia* with the schistosomes based on morphology. If *Griphobilharzia* is a schistosome, dioecy (separate males and females) evolved once. If it is a TBF, dioecy evolved twice: once in the turtle lineage and again in birds and mammals. If *Griphobilharzia* is a schistosome, it would have infected the common ancestor of crocodiles + birds + mammals. If it is a TBF, *Griphobilharzia* would have had to make a secondary jump from turtles to crocodiles! Two evolutionary leaps are fewer than four; therefore, Occam's razor favors the placement of *Griphobilharzia* with the schistosomes on morphological grounds, not the TBFs. At the moment, molecular data takes precedence over morphological, and this strange worm is in the middle of a tug-of-war with the gene jockeys holding the upper hand.

For the next few years, references to *Griphobilharzia* popped up in various publications for myriad reasons, none aimed at settling its contentious evolutionary relationships. In 2010 I received a request to review a paper for *Vestnik Zoologii*, an English language journal published in Ukraine. The article was titled "A revision of species *Griphobilharzia amoena* Platt and Blair, Purdue et Melville, 1991, a parasite of the crocodile" by a group from Uzbekistan, headed by D. A. Azimov. I was vaguely familiar with Azimov's work. I reviewed another paper of his for the same journal earlier in the year and read other articles he published, although only peripherally related to my interests. I had no opinion about him one way or the other.

Azimov's paper was like getting hit between the eyes with a 2 × 4! He and his cadre argued I completely misinterpreted the morphology of *Griphobilharzia,* and it wasn't dioecious after all. They claimed it was a monoecious worm belonging to a well-established genus of TBFs, *Vasotrema,* described by Horace Stunkard in 1928. After reading the manuscript, I was furious! I needed to talk to someone, not any someone, a parasitologist: someone knowledgeable, whose judgment I trusted. I called Eric Hoberg, who succeeded Ralph Lichtenfels as the curator of the USNPC. Eric and I met during our graduate school days. He is a top-flight taxonomist, although he works primarily on tapeworms and nematodes (roundworms) and does extensive work in the countries of the former Soviet Union. Eric had contacts there and might be able to give me some insight into Professor Azimov and his motivation for writing what I viewed as a load of crap. I called Eric semi-regularly for no other reason than to be able to talk with a kindred spirit.

To call Eric low-key is an understatement, at least when bad science isn't the subject. A friend nicknamed him "Eeyore" after the Winnie-the-Pooh character for their shared demeanor and sometimes dour outlook. Eric patiently listened as I outlined the thrust of the manuscript that set my hair on fire. When I finished, Eric paused — a little longer than usual and shared that he received the same paper to review. I was surprised, but there was more. Much more. As the curator of the USNPC, Eric's responsibilities included approving and arranging loans from the collection to scientists worldwide. Occasionally, folks requested permission to visit the facility and examine specimens onsite. Eric informed me Azimov and a few of his compatriots visited the museum and asked to see the slides of *Griphobilharzia* I deposited following publication of the original description. Eric obliged and reviewed the anatomy of the specimens with the visiting Uzbeks. He also provided photographs of various features of the specimens they requested. It was clear to Eric that Azimov and his cronies doubted the original description. Eric did his best to convince them they were wrong, not because we were friends, but because what he saw under the microscope agreed with the original description. The visitors left, returned to Uzbekistan, and wrote the paper they decided on before they traveled the 6,500 miles from Tashkent to Washington, D.C.

Eric and I agreed to write our reviews independently to mitigate any ethical issues. I cleared my desk, completed the review in a few days, and returned it to the editor. Eric did the same. Eventually, we compared notes, and we both recommended outright rejection. Our reasoning was similar, but the specifics differed. The editors agreed with our recommendation and declined to publish. After the rejection by *Vestnik Zoologii*, Professor Azimov had the temerity to ask me to join his group as a co-author and disavow my own work. I replied, informing him that I reviewed his manuscript, urged unequivocal rejection, and emphatically declined his invitation. I hoped we had slain the dragon. I was wrong.

A year later, a Google alert appeared in my inbox with the heading *Griphobilharzia amoena* accompanied by some Cyrillic letters. I clicked on the link and found that Azimov convinced a Russian language journal, *Parazitologiia*, to publish his nonsense. Once again, I went ballistic! I called Eric. He had seen the paper and was dumbfounded. I wanted to do something, but I didn't exactly know what that something might be. I have been wrong before. I have published things, later deemed incorrect, and somebody else set the record straight. But this was in a whole new galaxy. *I wasn't wrong!* The original description of *Griphobilharzia amoena* wasn't perfect. There were structures I wasn't able to document in as much detail as I would have liked, but what appeared in print was correct. Azimov and his Uzbek henchmen muddied the water with a paper that was wrong, wrong, wrong. I could not let it stand. I needed to write a rebuttal and conclusively demonstrate our original work was correct, and Azimov and friends were delusional. Before I could proceed, I needed to know precisely what the new paper said. A friend at Saint Mary's had worked as a translator for the Defense Department earlier in her career and was fluent in Russian, German, and French. I paid her to translate the Russian version. It was identical to the earlier iteration rejected by *Vestnik Zoologii*.

The *Journal of Parasitology* has a category of paper called *Critical Comments*. It had never, to my knowledge, been used to discredit an article from another journal. I wasn't sure how to proceed. The journal recently changed editors. Jerry Esch of Wake Forest University stepped down after a successful run of almost two decades and was succeeded by Mike

Sukhdeo of Rutgers University. Both were friends of long standing. If I were to pursue this course, Mike would be the final arbiter in the decision to accept or reject the manuscript. I wanted some guidance before I took the next step. I called Jerry and outlined the situation from the beginning.

I planned to relate the entire history of the episode, provide additional information supporting our original findings and discredit the group from Uzbekistan. Jerry listened patiently and indicated his support for my plan. I contacted Eric and asked if he would co-author the paper with me. I also approached Leslie Chisholm, Curator at the South Australia Museum, with the same offer. Azimov contacted Leslie and requested photographs of the type specimens of *Griphobilharzia* I deposited there as well. She had also seen the article in *Parazitologiia* and agreed it misrepresented the original description's factual basis. Leslie joined us as the third member of the team.

I started writing. I included a historical review, and Eric and Leslie sent copies of the photographs they supplied to the Uzbeks. I finished the initial draft and sent it to Eric and Leslie for their input. Both are incredibly able scientists and excellent writers. They improved the paper significantly. They were concerned with the level of vitriol I leveled at my critics and strongly suggested I tone down the rhetoric. I wrote some things in the heat of the moment that should never find their way into print. I removed everything they found objectionable. We assembled the photographs to accompany the article and submitted it for review.

We heard nothing for several months. Then the "shit" hit the fan! Mike Sukhdeo, the editor, was apoplectic. Editorial notes accompanying reviews were always maddeningly vague and noncommittal. Please read and respond to the reviewer's comments, blah, blah, blah. Not this time. Mike called the manuscript "horrible" because we (more accurately, I) attacked another scientist. Even though, in my mind, the scientist, Dr. Azimov, deserved our ire. A member of the editorial board talked Mike out of summarily rejecting the paper, and we should have the opportunity to revise it. One reviewer had little to say other than we should remove any personal invective, and the second reviewer's comments went on for pages.

Most research adds new information or insights to the scientific canon, the ultimate goal of all science. This paper was different. We were correcting an egregious example of promulgating something not only wrong but bad science. Mark Twain once wrote, "a lie can travel halfway around the world while the truth is putting on its shoes." I wanted to stop Azimov and his cadre in their tracks.

We answered our critics, and our paper entitled "On the Morphology and Taxonomy of *Griphobilharzia amoena* Platt and Blair, 1991 (Schistosomatoidea), a Dioecious Digenetic Trematode Parasite of the Freshwater Crocodile, *Crocodylus johnstoni*, in Australia" appeared in the October 2013 issue of the *Journal of Parasitology*,[66] approximately 18 months after publication of the Uzbek paper. Fortunately, at least in this case, foreign language papers tend to attract much less attention than those in English. *Griphobilharzia* = *Vasotrema* hadn't drawn much interest since it appeared. I sent Azimov a copy of our rejoinder and waited for a response. Silence. While I consider the matter settled, I keep waiting for a Google alert in my e-mail inbox. Now, if we can reconcile the molecules vs. morphology conundrum, I would be ecstatic!

12. Ten Months Away from Home

Wherever you go becomes a part of you somehow.

— Anita Desai

Our flight from Honolulu landed in Cairns in the early afternoon. We cleared immigration, customs and boarded a plane for Townsville. David Blair met us with a Toyota 4 × 4, an absolute beast of a vehicle with a 'roo bar (to protect the vehicle in the event of hitting a kangaroo) and the steering wheel on the wrong side of the cab. We loaded our suitcases; David tossed me the keys and told me to follow him. I looked at the keys and back at David. I never drove on the left side of the road before, but I figured if I hit anything smaller than a semi, I would win. So off we went.

I started the behemoth, put it in gear, and followed David to a caravan park. I quickly ascertained the basic rule of driving anywhere: the driver is always next to the centerline. Obey that imperative, and you will be fine. Our first 'home' in Australia was a trailer at a local campground on the edge of Townsville. We grabbed dinner at a nearby restaurant and settled in for the evening. I laid down, turned on the TV, and attempted to figure out the rules of cricket. I nodded off with no idea what was going on. House hunting would begin in the morning.

Kathy and I decided to focus our search based on two criteria: schools and proximity to James Cook University. We settled on the Heatly school district, about four miles from campus, and rented a three-bedroom highset at 381 Fulham Road. The main floor of a highset is about ten feet off the ground. The space below frequently serves as a garage and utility/storage area. In theory, the design improves air circulation and helps cool the home during the relentless summer heat — which was pretty much all year. It was just worse from December to May.

We needed a car. I couldn't keep the 4 × 4 forever. Kathy had one request — air conditioning. I failed. I got a good deal on a used Toyota sedan with 116,000 kilometers for $5,000, but no air. It was a great car, and I sold it for the same five grand when we left. We prowled secondhand stores for furniture, appliances, and nicknacks to make the place closer to a home than a house; however, the guiding principle was functional and cheap, with style a distant third. Next on the list — bicycles. We needed three: one each for David and Reid and one for me. Kathy would have the car, and I would commute to James Cook under pedal power. I estimated I biked about 2,000 miles during our ten-month stay. During the first two weeks, we furnished our home, got the kids enrolled in school, and I started organizing my office at the university. Our life in Australia was taking shape.

David Blair and his partner Dinah invited us to join them for some evening adventures bar hopping to establishments featuring local musical groups. Kathy is much more outgoing than I am, and she made contacts that would prove propitious in both the near and long term. Australians are friendly and view Americans as cousins. British convicts that eventually populated the country were sent to Botany Bay in the late 1700s as the nascent United States closed their access to penal colonies in Georgia. Also, Australians fought as allies of the United States in every war from World War I forward.

Our sons, David and Reid, made friends quickly as their new schoolmates were fascinated with our culture, and they were the only Americans around. Kathy even found a job as a volunteer (she could not work for pay) cataloging coral at the Museum of Tropical Queensland. Her stint in the biological realm was short-lived once somebody learned she had a background in fundraising.

We made friends, or I should say Kathy made friends and dragged me along. Early in our stay, Kathy met some outdoor types who invited her for a wilderness camping trip to Pelorus Island, a few miles off the coast a bit north of Townsville. The island had only one inhabitant who, for a fee, ferried groups to and from the camping area. The camping rules were simple; bring everything you needed during your stay, including water, and take everything with you when you left. The rules permitted poop burial; however, butt wipes went home.

Kathy met a couple from New Zealand, Steve and Roz. They were an interesting pair; he was a chimney sweep and an artificial inseminator of cattle while she sewed sails for the boats dotting the harbor in Whangarei. They lived in a cooperative on the North Island, approximately two hours north of Auckland. Kathy shared that we planned to visit New Zealand on our way home, and they invited us to spend our time with them. All I could say was, "Sweet!"

Days turned into weeks and weeks to months. We all settled into our routines. The kids went to school and hung out with their new friends. Bicycles, flat terrain, and fine weather provided a good deal of freedom for them once school let out. On weekends, trips to the golf course were the norm; however, the beach was off-limits due to *Chironex fleckeri*, a deadly box jellyfish inhabiting the waters of Northern Australia during the austral summer. Kathy spent time at the Museum brainstorming ideas for fundraisers, having lunch with new friends, or exploring the area. I was in the lab. We were preparing for family visits, Kathy's parents in mid-October, and her sister's family near the first of the year.

Going to a foreign country for an extended period almost guarantees elevation to tour guide status at some point. Kathy's parents, Bill and Linda Fenton, visited for ten days in October. We toured Paluma, the southern edge of the dry rainforest north of Townsville. We drove to Charters Towers, an hour and a half west of Townsville, dipping our toes into the Outback. Friends explained the difference between the colonization of the United States and Australia: we landed on the coast, crossed the mountains, and found the immensely fertile Great Plains; they did the same and discovered a desert. Once you travel an hour into the Outback, you can stop. The scenery doesn't change for the next 3,000 miles. The monotony of the region was captured on my favorite bar coaster. One side read, "What do folks do for fun in the Outback? [Turn over.]" On the reverse, "What do folks do for fun in the Outback? [Turn over.]" Not hard to see the irony, is it?

We took Kathy's parents to the Great Barrier Reef, a two-hour ride in a fast catamaran. Kathy's dad recently turned 80 and was beginning to show his age. Despite Linda's protests, Bill donned fins, mask, and snorkel, and aided by a life jacket, explored the beauty of the reef firsthand. He rated the experience as one of the most memorable of his long life. At the end of their visit, Bill and Linda left for a few days in Brisbane, a few more in New Zealand, and home via Japan. Kathy's mother took a short trip to Cuba before Castro established his worker's paradise. Besides brief vacations to Canada, I don't think either of them had been out of the country. We did our best to make their big adventure as memorable as possible.

The Brininngers, Kathy's sister's family, joined us shortly after Christmas. Pat, Chuck, Julie, and Todd arrived from Jamestown, New York, trading the winter's depths for a steaming tropical summer. They left snow and temperatures hovering near 0°F and arrived to find triple digits almost every day in Townsville. We repeated many of the same activities we shared with Kathy's folks and headed south for the slightly less oppressive temperatures in Sydney. We did the usual sightseeing and took the opportunity to attend one of the few Unitarian churches in the country. Locating it proved problematic as the address was identical to the Marriott Hotel. The church occupied some prime real estate coveted by developers. The congregation agreed to sell if the developers incorporated a church as part of the new building. We walked around the block until we noticed the "Flaming Chalice," a symbol of our denomination, above one of the doors nestled in the side of the hotel.

The Reverend Charles Eddis, a guest minister from Canada, conducted the service. We chatted up the folks sitting next to us only to discover my neighbor was the minister's wife. And things got strange, not weird strange, unlikely coincidence strange. It turned out Nancy Eddis served as the secretary for the Institute of Parasitology at McGill University in Montreal for many years before retiring. We chatted, before and after the service, about folks we both knew. It is a small world.

Kathy's family departed for home, and we headed for Tasmania. Wildfires ravage southeastern Australia almost every year. A group of Tasmanian firefighters shared our flight after doing their bit for their northern neighbors. They were welcomed back as heroes, which they were. A group of their comrades doused our plane with water cannons as we approached our gate — a tribute that delighted everyone onboard.

We completed the checklist of tourist outings: cruised the forests and noticed many Tasmanian devils killed by encounters with vehicles, drove to the top of Mount Wellington for a truly spectacular view of Hobart and the harbor, and an obligatory visit to the Tasmanian Museum. I employ one of two strategies when museum gazing: 1) make a quick pass taking in as much as possible in a short period sacrificing the details, or 2) spend time carefully reading each panel in a small section and miss most of what's on offer due to lack of time or exhaustion. I chose the second option.

My focus was the deplorable treatment of the aborigines by white settlers. I was impressed by the forthright depiction of the cruelty the natives endured. Aborigines were considered vermin to be hunted for bounty and exterminated. When confronted with regulations for the humane treatment of animals in research, I often think about how low the bar "humane" is when it comes to our fellow human beings and how high it is for research animals.

Our final stop was the Gold Coast, the Miami Beach of Australia, located south of Brisbane. The beaches are south of the Great Barrier Reef and not protected from offshore winds and the rough water they generate. Unfortunately, the weather was less than congenial on all but one day. The beaches were crowded but closed for swimming for most of our visit. However, Reid and David's enjoyment of their time at the shore was undiminished. Friends told us any beach in Australia could be a topless beach. Swimming takes a backseat to nudity for 12- and 15-year-old boys. They were incredibly circumspect. Our boys lay on their towels, feigning sleep while surreptitiously casting their gaze on young women sunbathing in monokinis. They didn't leer, point, or otherwise call attention to themselves. I couldn't fault them for looking at what many young women offered to the public.

We returned to Townsville. I went back to the lab. The kids went back to school. And Kathy continued planning the museum fundraiser scheduled for the week before we returned home. The last few months of our stay were routine, with two exceptions: a visit to the Atherton Tablelands and Kathy's fundraiser.

Before we departed for Australia, Kathy met two Aussies on sabbatical at Notre Dame. She asked if they could visit one place in their home country, what would it be? Some well-known destinations such as Uluru (or Ayers Rock), the Great Barrier Reef, and Sydney Harbor come to mind, but their suggestion left us scratching our heads — the Atherton Tablelands. We had no idea what to expect when we headed north for the five-hour drive from Townsville to this can't-miss destination.

We arrived at the caravan park and unpacked. The trailer was a replica of where we spent our first nights in Australia. For the next few days, we drove and took in local attractions. The countryside was picturesque: green rolling hills and dry tropical forest. The flora and fauna were distinctly Australian, but what struck us was how much it looked like northeastern Ohio where Kathy and I grew up, and then, an epiphany! Australia is mostly desert. Mile upon mile of flat, arid land painted in hues of brown and red. Lush, green vegetation is rare. A scene out of the ordinary. A feast for the eye. To Aussies, the Tablelands are unique and something worth going out of your way to see. Little did those well-meaning folks realize they were sending us home.

Kathy and I walked after supper, an irregular habit. She talked, and I listened and grunted or nodded at what I perceived to be appropriate moments. She needed to be heard, taken seriously, and to have what she believed was my undivided attention. All of this happened when we walked, and I did my best. During these strolls, our conversations focused on our children: their progress in school and behavior. As we neared the end of one meander through the neighborhood, she asked an interesting question. What would I miss most after we were back in South Bend? She didn't like my answer or the speed of the reply. My response? Nothing.

There would be a few fond memories. I was sure I would reminisce and tell stories of crocodiles, turtle hunts, and snorkeling the Great Barrier Reef. Her question implied a sense of longing — a desire to recapture something lost if only you returned. And the answer was nothing. I could not conceive of circumstances that would bring me back to Australia not involving work: a functional solution to a metaphysical question. I would come back to work. I would not return because there was something I wanted to recapture. There was nothing to miss.

Kathy sold the museum folks on an art auction as a fundraiser with the theme of *Champagne and Chocolate*. She compiled a list of local wildlife artists and approached them about donating a piece to benefit the museum. Most had a reserve price which they would receive, and anything above the reserve went to the museum. Kathy organized the food, drinks, and venue. She transforms into a different person when she is in the zone. The sweet and genial woman I love becomes a single-minded juggernaut. She knows what she wants and expects everyone to follow her direction. If you can't, or won't, she will do it herself. She can be demanding and short — not short-tempered; she rarely gets mad. I do my best to follow directions and stay out of her way.

Once the auction started, Kathy relaxed and enjoyed the fruits of her labor. It was a wonderful evening. Everyone had a marvelous time, and compliments made up for the time and energy Kathy poured into the event. The net result was the most successful money-raising affair in the museum's history — a bit over $8,000!

I love having original art in our home and thought a painting representative of Australia would be a fitting souvenir of our visit. I had my eye on a lovely painting of a cassowary guarding her nest containing five green eggs. The bidding quickly went to $100, $150, and $200. At that point, there were only two bidders, and I was determined to win. Bids crept slowly higher, and I raised my hand when the auctioneer called out $350. I felt a hand on my shoulder. One of our friends begged me to stop bidding. I looked at him quizzically. He shared our friends were trying to buy the painting for us as a "surprise" going-away gift, and I was screwing up the plan, not to mention making their gift ever more expensive. I put my hand down; the painting went for $400. Kathy and I were touched by the thoughtfulness and generosity of those folks who had become dear friends in such a short time. Their gift is a treasured reminder of our Australian adventure.

Our final week focused on divesting ourselves of the odd collection of stuff acquired over the past ten months and consolidating the remainder to the original seven suitcases plus golf clubs we arrived with. I boxed the books, papers, specimens, and other odds-and-ends from the lab to ship home by the cheapest means possible. We started cleaning to recoup our

rental deposit. We all pitched in scrubbing, sweeping, and dusting in anticipation of the review that would determine the fate of our security deposit.

Colleen, the rental inspector, felt the standard for compliance was a
clean room in an operating theater or computer chip facility. She found
our efforts wanting. Colleen dragged her finger along the tracks of the sliding windows in the front of the house, and it came back dusty. Her disdain
was palpable. The owner was an "arse" and lied about the place's condition
when we moved in. He claimed we should have had the drapes dry cleaned
before vacating his property. Fortunately, I had the original inspection
report stating the curtains were dusty and full of cobwebs when we arrived.
We won more than we lost and got most of our money back.

Ten months was more than enough time to change hearts and minds.
As we were waiting to board our plane in Chicago, I half thought David
and Reid would try to convince airport officials they were being kid-
napped to avoid leaving. Ten months later, it was the same — neither
wanted to return home. David had a wonderful group of friends, including
his first girlfriend, who desperately wanted him to stay. Reid's basketball
coach chose him for an all-star team set to participate in the Queensland
State Tournament. Going home was just as hard, in a different way, as leaving in the first place.

We made good friends during our stay. They included us in family
gatherings in their homes, which were much nicer than our rented digs on
Fulham Road. A bit of a "mob" appeared at the airport. A dozen friends of
our kids were there to say goodbye, as were a similar number of friends
from the "Uni" and elsewhere. Our new friends' warmth and generosity
made me reassess my earlier "nothing" response to Kathy. I would enjoy
the opportunity to rekindle those relationships. We flew to Brisbane,
Sydney, and on to Auckland for a five-day visit to the commune near the
sailing port of Whangarei as guests of Steve and Roz, whom Kathy met
during our first week in Townsville.

Steve made the 200-mile round trip to pick us up and return us to the
airport. The only thing he and Roz asked in return was we purchase as
much duty-free booze as legally permitted. Their house was simple, with
interior lights powered by a car battery and solar panels to heat the water.
A bathtub and shower graced the front yard facing a vast and verdant

valley. Seven-foot screens were strategically placed to provide privacy from the house but left an unobstructed view of everything else. The 'loo was the most elegant outhouse I ever encountered. It was tastefully decorated, and I didn't have to sit on my hands.

Steve and Roz hosted a "Hunter-Gatherer" party in our honor. These folks love to party, and we were a convenient excuse. The rules were simple: only bring something you hunted, gathered, or raised yourself. Nothing from a store, although apparently, the whiskey we bought was exempt. Early the next morning, our band of six set off for some hunting and gathering. We stopped by the side of the road near a fenced pasture, and Steve pointed out mushrooms dotting the grassy slope. I knew something about mushrooms from my days at Hiram under the tutelage of Dr. Berg, and more than a few are quite toxic. I raised my concerns with our hosts, but Steve assured me these were safe to eat. I took his word for it. We filled several bags with fungi, got back in the car, and headed down the road.

We turned off the asphalt onto a gravel drive stopping at a locked gate at the bottom of a large hill. Steve and I were going snorkeling, and I could only assume the ocean was beyond our view. We unloaded wetsuits, masks, fins, snorkels, and mesh bags for collecting the lobster and abalone Steve assured me were available on the other side of our climb. We gathered our supplies and headed up. Upon reaching the crest, our little band had a spectacular view of a secluded bay leading to the Pacific Ocean. Whangarei is as far south of the equator as the northern border of the United States is north. It was late fall in the southern hemisphere, and the water would be cold. Wetsuits were required to avoid hypothermia. It was not the Great Barrier Reef. Steve and I donned our suits, and he instructed me on how to collect abalone.

The water was clear and cold. Giant kelp, reaching from the bottom to the surface, was the dominant vegetation. It was stunning, beautiful in a way utterly different from the tropics. The colors were muted browns and greens, and the light danced and shimmered, casting shadows with the swaying movement of the kelp. I got to work on the abalone while Steve headed out to deeper water in search of lobster. An hour and a half later, I had a dozen abalone ranging in size from 8–10 inches, and Steve managed to procure four very impressive lobsters. Despite the protection of the wetsuit, I needed about 15 minutes to stop shivering once I was dry and fully clothed. We repeated the climb, got in the car, and returned to the house to prepare for the evening's festivities.

Guests began arriving around 7 pm with their offerings from gardens, farms, fields, and sea. They also brought music — guitars, violins, and other instruments in various shapes and sizes. It was a raucous evening. The duty-free liquor disappeared quickly. Wine bottles appeared next, followed by the sweet smell of marijuana permeating the room. Songs, some familiar and others not, rose and fell through the ensuing hours. We sang and danced and danced and sang. The last of the hunter-gatherers left around 3 am.

During the evening, Steve, an aspiring musician, and a neighbor, Renee, who had a home recording studio, agreed to meet the following day for a recording session. Steve was a more than competent guitarist but lacked the confidence to sing with authority. Renee suggested a backup singer as a solution to Steve's reedy voice. No hands went up. Kathy has a lovely voice but hadn't sung except for hymns during church in years. After a few minutes of cajoling, Kathy was in the studio wearing headphones, standing in front of a microphone outfitted with a disperser. Steve and Kathy did a trial run while Renee worked the soundboard. After a few takes, Renee had what he needed. He added a bass track, and after a bit of mixing, we heard a beautifully produced piece of music. I was so proud of Kathy words failed me.

The following day, we said our goodbyes, and Steve drove us back to Aukland for our return to the States. We crossed the International Date Line regaining the day we lost the previous year. 16 hours later, we landed in Chicago, retrieved our minivan, and drove back to South Bend. Our house was still standing and in reasonable condition. However, I am sure Colleen would have found any number of housekeeping transgressions warranting confiscation of most, if not all, of our renter's security deposit. None of us cared. We were home.

13. Australian Research

This is the real secret of life — to be completely engaged with what you are doing in the here and now. And instead of calling it work, realize it is play.

— Alan Watts

Australia was more productive than I could have hoped. Our nearly year-long excursion resulted in 16 publications over the next decade and a half. The first paper came before I left the United States and paved the way for the trip and everything that followed. Unraveling the mystery of *Griphobilharzia amoena* (Chapter 11) from the freshwater crocodile forged my relationship with David Blair. His support was essential for the success of my grant application to the Lilly Endowment. My association with David led to a collaboration with Sylvie Pichelin and Tom Cribb and the description of the first TBF from Down Under. Tom felt comfortable entrusting me with specimens from an unpublished thesis, which resulted in the description of three new genera and a slew of new species parasitizing Australian turtles. The first falling domino set in motion work that occupied me into the new millennium.

I had two objectives for my time in Australia: 1) learn the application of molecular biology to trematode systematics, and 2) dissect as many turtles as my permits allowed with the hope of finding new TBFs. During the first month, I got to know the area and wrangle introductions to anyone who could help me catch turtles. Seining was effective in shallow water with low visibility and snorkeling when the water was clear. I employed both with success. Seining requires at least two people rapidly dragging a net strung between two poles through the water and onto shore. Snorkeling is, well, a lot of fun.

My first group of turtles came to an untimely and unexpected end. I had nowhere to house live turtles at the university, so I commandeered the washtub below our highset. One afternoon I came home from the lab, and Kathy was visibly agitated. In her usual style of conveying bad news, she hemmed and hawed and danced around the subject until I asked her to "please" tell me what was wrong. "Your turtles are dead!" she replied as tears welled in her eyes. Kathy decided to do laundry, and the hot water from the machine backed up into the tub and killed my animals!

I couldn't understand how it happened. The washing machine emptied into a joint discharge pipe with the washtub and to a culvert spilling into the storm drain. The drain pipe should have carried the water into the sewer system bypassing the tub. I removed the turtles and located the source of the problem. A frog took refuge in the drain pipe. When the hot water hit the frog, it died, forming a plug preventing the water from exiting to the culvert. The hot water backed up into the tub, parboiling my turtles. I hugged Kathy and assured her it wasn't her fault. Murphy was so right, "Whatever can go wrong, will."

I snorkeled for turtles at Paluma Dam, a reservoir north of Townsville, and Alligator Creek, about a half-hour south. Despite its name, Alligator Creek is not home to crocodilians of any sort, at least not the stretch I was exploring. Catching turtles underwater is not as easy as it might seem. They are wary and swim faster than most folks imagine. I had to approach them with as much stealth as possible and hope to maintain the element of surprise before they dove for deep water.

On one occasion, I was cruising along the surface of a deep pool of slow-moving water at Alligator Creek. I spotted a turtle resting quietly on a pile of vegetation about five feet below me and ten yards away. I began my descent, slowly angling down in the direction of my prey. I was sure s/he would sense my approach and bolt for the depths. I was within an arm's length, but the animal remained still as I reached out and grabbed it. Two things happened. My quarry began struggling to escape, and the pile of vegetation exploded as a second turtle bolted for deeper water. My catch was a male, and I surmise he was waiting for the opportunity to mate with the female hidden from view. He fell victim to one of the Seven Deadly Sins — Lust!

I was busy doing necropsies, collecting worms, and making slides. Once the slides were ready, I started the process of identification. David had an impressive collection of keys, which he happily let me borrow. Combined with the literature shipped from home, I had all I needed to get started. I quickly identified *Sigmapera cincta,* an intestinal trematode described in 1918 by William Nicoll, a British parasitologist. I was now confident about identifying the remaining specimens. I was wrong.

I knew from experience several worms belonged to several distinct genera, but I had no luck in bringing their identities to light. Keys are imperfect instruments because their authors are imperfect humans. I frequently arrived at an impossible answer, went back, and tried another path. If I failed to reach a satisfactory result, I looked at the pictures until I stumbled onto something similar to the specimen in question. The illustrations failed me as well. Sherlock Holmes said it best, "When you have eliminated the impossible, whatever remains, however improbable, must be the truth." The truth was these worms were new. I was ecstatic.

One morning, I shared my enthusiasm with David over a cup of tea for him and coffee for me. With his typical laconic Scottish demeanor, David looked up from his drink. He shared that he thought a graduate student from the University of Queensland described a bunch of new turtle trematodes for his PhD but never published his work. Maybe I should check it out. I returned to my office and immediately e-mailed Tom Cribb, the parasitologist at UQ, inquiring about this unpublished work and if he might send me a copy. David's recollection was accurate, and Tom responded promptly with a photocopy of the thesis.

The joy of discovery was short-lived. Everything I couldn't identify by reference to the published literature was right there in an unpublished thesis. The author's name? Lindsay Jue Sue. After completing his PhD with Tom's predecessor, John Pearson, in the early 1970s, Lindsay left UQ without publishing his work. I wasn't sure what to do. A thesis doesn't qualify as a formal publication for establishing new names of organisms, so I couldn't reference Lindsay's work to report what I found. I could describe them myself; however, I faced an ethical dilemma as Lindsay found worms a decade and a half before I did. I wasn't sure what to do, but a solution would present itself in a few months.

Tom Cribb and Sylvie Pichelin were strangers when Sylvie's letter arrived in South Bend in 1992. Publication of *Griphobilharzia amoena* the previous year and several other papers on TBFs elevated me to an expert on the group and worthy of consideration as a collaborator. Sylvie collected specimens she correctly suspected as TBFs and offered to send them to me. I would describe them, and we would share authorship on the completed paper. Once I met Sylvie and Tom and learned both were talented taxonomists, I wasn't sure why they felt they needed my help. She sent the specimens, and I got to work.

By the time I arrived in Townsville the following August, I had a preliminary manuscript in hand. Sylvie and I presented the results at the Australian Society of Parasitologists meeting held on Heron Island at the south end of the Great Barrier Reef in September. We decided to submit the paper to *Systematic Parasitology*, a British journal, edited by David Gibson of the Natural History Museum in London. I took the edits Sylvie suggested, returned to Townsville, and retyped the paper. I made copies and dropped the envelope in the mail for its trip to the British Isles. I went back to work and didn't think much about the paper for the next couple of months.

The reviews finally arrived and were generally positive. I dealt with the criticisms quickly, save one. David Gibson, the editor, wanted me to change the spelling of the family name of the TBFs: Spirorchidae to Spirorchiidae. Horace Stunkard established the family of TBFs in 1921 under the name Spirorchidae. In the world of nomenclature, priority is nearly sacrosanct. The first name appearing in the literature prevails.

The identification of new taxa is a biological issue; the formation of their names falls in the realm of nomenclature. The International Code of Zoological Nomenclature (ICZN) details the construction of names. It is a tome covering several hundred pages (in English and French) that would leave the most astute lawyer scratching their head.

The variant spellings of the family name of the TBFs, 'i' versus 'ii,' plagued the literature for decades; some workers used one, while others used two. Several years earlier, the editor of the *Journal of Parasitology*, Brent Nickol, asked me to resolve the issue as part of a paper I submitted describing a new genus and species in the family.[24] I consulted classics scholars about the root word -orchis (testis) and how the original generic name, *Spirorchis* (spir- for spiral and -orchis for testis), should be used to accept the mandated suffix, -idae, the required designation of a family in zoological nomenclature. The unanimous opinion of these folks was an additional 'i' should not be attached to the root 'Spirorch-' when adding the suffix "-idae" and the correct spelling of the family name was Spirorchidae. David was adamant. He demanded Spirorchiidae. Other families of trematodes with the suffix '-orchis' used the 'ii' construction, and he wanted conformity in the spelling of family names. I disagreed.

We exchanged several e-mails arguing the point, and neither of us blinked. David wouldn't publish the paper unless I changed the spelling, and I wouldn't acquiesce. Depending on your point of view, either he rejected the manuscript, or I withdrew it. I submitted it to the *Journal of Parasitology* and endured a second review process. The paper was accepted and published with one 'i' in the family name.[30] Our impasse supports the old adage, "the fights in academics are so vicious because the stakes are so small."

David and I rehashed our disagreement almost a decade later. He approached me to contribute a key to the genera of TBFs[43] as part of a multi-volume effort to provide an identification guide to all the genera of trematodes known at the time. David, as an editor, insisted on his pre-ferred spelling Spirorchiidae and I renewed my objections. We compromised. He used his spelling in the text, but I was permitted to include a footnote presenting the case for Spirorchidae. He agreed, and the reference appeared at the bottom of the first page of the chapter. In the end, David won. More and more people chose consistency over what I still view as correct.

My interest in collecting and examining turtles extended beyond the confines of Queensland and Townsville. I arranged collecting trips to northern New South Wales, southern Queensland, and Western Australia. Finding new TBFs meant examining as many species of turtles as possible, which meant traveling.

I spent the first ten days of November at the University of New England as the guest of Klaus Rohde, a German transplant who held various positions in Germany and Southeast Asia before landing in Australia.

Klaus was a genial host, but we had little interaction as he was busy with classes and his own research, and I was either in the field collecting or in the lab dissecting turtles. The two most common turtles in the area were a species of long-neck, *Chelodina longicollis*, and an undescribed species of *Emydura*, a close relative of *Emydura krefftii*, so prevalent in the waters near Townsville.

The long-necks were common in the dams (Australian for an artificial pond) scattered on the farms in the area. *Emydura* sp. was another matter. I needed assistance catching both. Wayne Higgins, Davey Dye, and Mick Cornish were employed by the University in various capacities and somehow drew duty as my assistants. They were, in local parlance, "good mates."

The first few days revolved around *Chelodina longicollis*. The 'mates' and I drove about 30 kilometers outside Armidale to Mick's father-in-law's farm and snagged about a dozen unlucky long-necks from several of the dams on the property. They yielded an assortment of worms, nothing not easily recognized but no TBFs.

Later in the week, we left for Smith's Creek, a tributary of the Macleay River on the coast of northern New South Wales. We were in search of the unnamed species of *Emydura*. The drive was spectacular. Most of the trip was on a logging road, often no more than a single lane that dropped over 3,000 feet from the escarpment to the coast. The scenery was nothing short of breathtaking, and the ride nerve-racking. Wayne drove out, and Mick drove back. Neither wasted any time getting where they were going. I noted one car that failed to negotiate a turn and landed on its side, balanced atop several trees 100 feet below the road, which left me wondering about the occupants' fate of that strangely perched vehicle.

Wayne talked with the owners of the property adjacent to the site we wanted to explore, and they were more than happy to allow us access to the river. Wayne and Mick were both divers and looked forward to spending time in the water, even if only a few meters deep. The initial results were not encouraging. Wayne came back after 30 minutes with one turtle and a bag of mussels — good eating, he said. The next pass was more successful. Mick emerged with one, and Wayne had seven. Wayne did another ten minutes and returned with the last specimen. The resulting necropsies were the same; nothing unfamiliar and no TBFs. Disappointment didn't come close to what I felt. The bottom line? I spent ten days, $1,000, and found no TBFs — new or known.

In mid-December, I flew to Perth and the Murdoch School of Veterinary and Life Sciences as the guest of Russ Hobbs, a graduate school chum from the University of Alberta, and his partner, Sue Harrington. Russ and I shared the same graduate advisor, Bill Samuel. Russ did his Master's thesis on a complex community of nematodes in the intestine of pika, a small rodent inhabiting the rock scrabble on northern mountainsides worldwide. Russ's thesis was, to put it mildly, a masterpiece. Rumor had it, after his thesis defense, his committee offered to award him a PhD instead of a Master's. Russ declined. He stated he came to UA for a Master's, and that was what he earned. If true and not apocryphal, Russ was insane. I never asked. It was better not to know.

My flight landed, and Russ took me directly to Murdoch to check traps he set earlier. We found four *Chelodina oblonga*, one of the western species of long-neck turtles — a great start. After returning the turtles to the lab and ensuring their safety until the next morning, we headed to Russ' house in North Fremantle, which guards the entrance to the harbor serving Perth, the West's largest city.

The Hobbs/Harrington home was a small, single-story affair dating to the early 1900s. The original house consisted of four rooms, all with beautiful wood floors and molding, and tastefully decorated. The back porch was enclosed and converted into a kitchen and bathroom. Paintings adorned the walls, primarily watercolors, several painted by my host. Sue and Russ were non-militant (no plastic explosives or Molotov cocktails lying about) ecowarriors: total recycling, composting in the backyard, minimal driving, no television, lots of vegetarian fare, no dessert, and invigorating ocean swims at 6 am. I am nothing if not the perfect guest. I did it all without complaint — and enjoyed most of it.

Turtles were in short supply, so Russ and Sue suggested a tour of the area. The British established Perth as a western port for their burgeoning presence in Australia. Where there are sailors, there will be a gaol (British for jail). The Roundhouse, a prison that could have had an ocean view if the Brits had thought to put windows in the cells, dominated the harbor. I found it interesting so many tourists spend time and money to view sites of such human misery and suffering. They peered in the cells and uttered nostrums like, "Can't imagine being cooped up in there." Really?

We set four traps per night for six nights and caught five turtles, only one more after the first day. None of them were infected with TBFs — c'est la vie. I did discover much later there were two new species of trematodes in our limited haul. I cut my stay short and took the "midnight horror" (= red eye) home.

In mid-February, I flew to Brisbane for turtle collecting as the guest of Tom Cribb and Sylvie Pichelin. My goal? Collect turtles from the same spot where Sylvie found the only TBF known from Australia at the time. It seemed the best bet to obtain new material for both morphological and molecular studies.

Tom, Sylvie, and I packed a vehicle with all the gear needed to trap turtles and headed southwest toward Warwick. Our objective was Leslie Dam. We stayed at the nearest and cheapest motel we could find — The Buckaroo Motor Inn. We checked in and drove to the dam. The lake was a barren body of water. Two pelicans and a few gulls gliding lazily overhead comprised the only visible wildlife. I hoped there were some turtles hungry enough to find their way into traps baited with ox hearts. Water to fill the lake came from tropical storms, and there hadn't been one in years. The distance from the spot where Sylvie set her traps several years earlier to the current water's edge was maybe 50 meters. In the absence of a substantial downpour, Leslie Dam might well disappear.

The following day was ugly: dark skies and a persistent drizzle as we headed back to check the traps. By the time we arrived, the rain had abated. A lone truck and boat trailer graced the parking lot, but there was no boat in sight. The first trap yielded two *Emydura macquarii*, the remaining three only one, and the last one lured five, so I was two short of my quota. The rain stopped, and Tom decided to throw in a line. Sylvie and I walked down to a rock outcropping to search for snails. Tom caught a turtle with his fishing gear, and there was one more in one of the traps, so we packed up and headed back to the motel. After a quick shower and a bit of rearranging of junk in the back of the troop carrier, we were on our way back to Brisbane in a steady and more substantial rain.

The turtles of Leslie Dam did not disappoint. I finally saw my first living Australian TBF — *Uterotrema australospinosa* (Figure 5). I took my time and admired the living specimens under the dissecting microscope. They were beautiful and exactly (at least at low magnification) as described

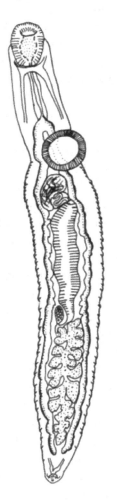

Figure 5.   *Uterotrema australospinosa* from: Platt, T.R. and S. Pichelin. 1994. *Uterotrema australispinosa* n. gen, n. sp. (Digenea: Spirorchidae), a parasite of freshwater turtle *Emydura macquarii* from southern Queensland, Australia. *Journal of Parasitology* 80: 1008–1011. (Reprinted with permission.)

in the paper we submitted to the *Journal of Parasitology*. Sylvie asked me if I noticed the fluorescent nature of the uterus. I looked back into the microscope, and I saw what she meant. With sub-stage illumination, the cells lining the uterus appeared bright red; change the background to black, shine the light from above, and it glowed fluorescent green. There must be an intriguing chemical in those cells to react so eerily. We included the observation in our paper but could not explain its utility to the worm. The turtles had many other familiar worms: various nematodes and trematodes. For all the time and expense involved in my exploration of the continent, my reward was two and a half TBFs. I began to question my decision to spend a year of my life and all it entailed for such a paltry return.

John Pearson, Tom's predecessor and mentor, visited the lab late one afternoon. John struck me as cultured and genteel. We chatted about various topics: principally TBFs and the spelling of Spirorchidae (one "i" or two). Fortunately, the issue of Dan Brooks didn't arise, or I might have had to do battle with a more-than-worthy adversary. John hated Dan with a red-hot passion. Dan published a long paper proposing a new evolutionary scheme of all of the major groups of trematodes. John disagreed, to put it mildly, with the results, and spent several years working on a treatise to discredit Dan's work. John's opposition was deep and unwavering. I kept my relationship with Dan to myself and stayed on safe ground. We parted on good terms, or so I thought.

Most of the trematodes I found in the gastrointestinal tract of Australian turtles were new to science. However, it didn't mean they were unknown. Lindsay Jue Sue produced a beautiful thesis with detailed accounts of the life cycle of each new species. Since Lindsay never published his results outside his thesis, they didn't exist. During my visit, Tom made me an offer I couldn't refuse. Take all Lindsay's specimens, and get the work in print. The only conditions were 1) accord Lindsay the honor of first author, and 2) thank Tom in the acknowledgments. I took it. Why would Tom give up such a seemingly straightforward project? He was in the early stages of his career, and his primary interest was parasites of fish from the Great Barrier Reef. As Tom explained it at the time, he had about eight projects in progress, and the turtle stuff was at the bottom of the list. He was glad to hand them off to someone else.

I started with *Sigmapera cincta*, a trematode described by William Nicoll, a British parasitologist in the early 20th century. According to David Gibson, Nicoll was a prolific author, iconoclast, and possibly a dipsomaniac (i.e., alcoholic). While more than adequate when first published, Nicoll's original description lacked important details and included no life cycle information. Lindsay's work would add significantly to our knowledge of this ubiquitous parasite.

My search of the literature revealed little new work impacting the information contained in Lindsay's thesis. I began reorganizing and editing the text. I knew how to write a taxonomic paper, so the writing was not difficult, but it took time. The illustrations were a real headache. Whether the description of a new taxon or the redescription of an existing form, any taxonomic work is accompanied by line drawings rendered in India ink. Authors frequently organize drawings into plates, with several figures per page. Each illustration is numbered and accompanied by a scale bar to indicate the size of the specimen, or structure, illustrated.

Lindsay's figures were beautifully rendered and appropriately organized for his thesis, but not for the papers I planned. I needed to cut and paste the figures into plates corresponding to the structure of the manuscript. Then I had to photograph the new compilations to the journal's specifications, develop the negatives, and print the resulting pictures on 8.5" × 11" photographic paper without any extraneous lines or markings — pure, bold black on a pristine white background. I was dealing with photocopies where the blacks were faded and not as crisp and sharp as the original. I am a self-trained photographer of modest skill. Achieving the exacting standards mandated by the journals took more time than I care to recall.

I submitted the paper to *Systematic Parasitology* for the simple reason they didn't levy page charges and provided 50 free reprints (copies of the article for the author to distribute in the era before the now ubiquitous pdf). Most people don't realize we pay to publish our research. Most journals don't carry advertising to support their operations, so they pick up the slack by charging authors a fee for each printed page. The other journals I considered charged between $50–100/page. Most researchers have grant money to cover these costs. I didn't. The first paper was about 15 pages, and I would be on the hook for $750–1500 out of pocket! Not to mention,

I was planning on three more papers of equal, or greater, length, a sum that would crush our household budget. Free sounded better.

I finished the paper and put it in the mail. Several months later, I received the reviews. The first reviewer suggested a few minor changes but nothing significant. The second review was similar: a few grammatical issues and typos. The third review was a whole different kettle of fish. Four pages, single-spaced, of incredibly detailed and stinging criticism. Most scientists don't sign their reviews. This one did, John Pearson, Lindsay Jue Sue's PhD advisor. I was stunned.

I spent the next few weeks going through John's critique and composing my response. He made a few valid points. I rejected another small group and offered a detailed explanation for why I was right and John was wrong. The vast majority fell into a third category. Why was something done or not done during the original study? I had no idea because I wasn't there, but he was. John supervised Lindsay's graduate work and affixed his signature to the finished thesis approving it as part of the requirement for awarding the PhD. I viewed Dr. Pearson's detailed, and in my mind, petty critique as nothing more than an act of petulance. I handled almost all of these in the same way. I merely stated I wasn't there, but the reviewer was, and he was in a better position to answer his own question. I finished my rebuttal, changed what I felt was necessary, and returned the manuscript to David Gibson. He accepted the revised paper for publication.[36]

I repeated the process three more times, and on two of the three, I received the same petty attacks from John. I fumed. I tracked down Lindsay Jue Sue and asked if he had any insights to explain his advisor's ill humor, but he was at a loss. He knew John wasn't happy he hadn't published his thesis work but felt they had a good relationship during his graduate studies. A little over 18 months after I started, the last article was in the hands of the publisher. The result was four papers totaling 73 pages: two new families, three new genera (*Dingularis*, *Choanocotyle*, and *Thrinascotrema*), six new species, and one redescription.[36–39]

A year later, I was examining my Australian specimens and reading the relevant literature. I stumbled across a paper by the aforementioned Dr. Nicoll in which he described a new genus and species of trematode from Australia in 1918 under the moniker *Aptorchis aequalis*. I looked at

Dr. Nicoll's drawings and compared them to specimens I knew as *Dingularis anfracticirrus*, one of the new genera and species described by Lindsay Jue Sue and now published in *Systematic Parasitology*. They looked the same, and my heart sank. We published a paper describing as new (the genus *Dingularis*) something already recognized and named *Aptorchis*!

One organism with two names may seem a small matter to most folks, but it isn't. Under the rules of Zoological Nomenclature, an organism can have one name and one name only. There were two issues. First, was the genus *Dingularis* the same as the genus *Aptorchis*? If so, one of the names had to go. The rules are simple. The first name applied has priority, and the second is considered a synonym and relegated to subordinate status. If I were correct, it would consign *Dingularis* to the scrapyard of zoological names as a junior subjective synonym. Confused? It means the name *Aptorchis* appeared first, and *Dingularis* second (hence junior to *Aptorchis*), and in my opinion (subjective), they represent the same biological entity (they are synonyms). Someone else could resurrect *Dingularis* if they could demonstrate I was wrong, and the two genera weren't the same.

Once I decided *Dingularis* was invalid, a second question remained: Were the species *Aptorchis aequalis* Nicoll, 1918 (Figure 6), and *Aptorchis anfracticirrus* (Jue Sue and Platt, 1999) the same thing? By the way, placing (Jue Sue and Platt, 1999) in parentheses means the species *anfracticirrus* was initially assigned to a different genus (i.e., the now-defunct *Dingularis*) and moved to the genus *Aptorchis*. Aren't you glad you asked? Again this determination is in the eye of the beholder, but I hoped to put the matter on solid ground by using a mathematical approach called Principal Component Analysis (PCA).

PCA converts the variability of each specimen's measurements into a single value and plots its position onto a two-dimensional graph. If all of the specimens cluster in one area, they are considered the same thing. If they form two, or more, separate groups there may be more than one species present. It is possible to determine where they are different and if those differences are biologically meaningful. All of the specimens clustered in one spot, strongly suggesting the three dozen trematodes collected from three distinct turtle species geographically separated by 1,500 miles

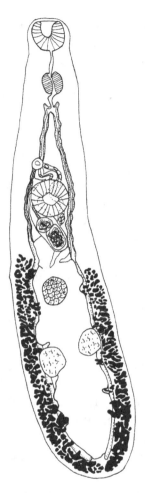

Figure 6. *Aptorchis aequalis* from: Platt, T.R. and R.J. Jensen. 2002. *Aptorchis aequalis* Nicoll, 1914 (Digenea: Plagiorchiidae) is a senior synonym of *Dingularis anfracticirrus* Jue Sue and Platt, 1999 (Digenea: Plagiorchiidae). *Systematic Parasitology* 52: 183–191. (Reprinted with permission.)

were all the same thing — *Aptorchis aequalis. Dingularis anfracticirrus* became a junior subjective synonym of *A. aequalis.* And another name bites the dust.[44]

This interlude represents one of the things I love about taxonomy. The ability to take specimens collected almost 100 years earlier, compare them to worms I collected, and determine with confidence they were the same species. The connection through time to our antecedents provides a powerful sense of belonging to an endeavor larger than yourself. It also makes you curious about the lives of your forebears. Who were they? What were their hopes and dreams? Did they have families? What happened to Dr. Nicoll? [Note: I've searched without success.] Will anyone think about me similarly a century from now?

I described several new taxa of trematodes from Australian turtles in the ensuing years. I named *Auriculotrema lechneri* n. g., n. sp. (new genus, new species) after Reggie Lechner, a dear friend and the former head of Maintenance at Saint Mary's.[45] I caught some flack from the parasitological community for naming *Buckarootrema goodmani* n.g., n. sp. after the Buckaroo Motel where Tom Cribb, Sylvie Pichelin, and I stayed on our collecting trip to Leslie Dam; and John Goodman, parasitologist and friend.[42] Some folks don't have a sense of humor when it comes to naming new organisms. I described two new species of TBFs and named one *Uterotrema burnsi* after my high school biology teacher, Alan Burns.[32] Mr. Burns was an inspiration.

I took on the task of clearing up some nomenclatural problems with TBFs of marine turtles and redescribed several species.[33,35] Finally, I described two new species in the genus *Choanocotyle*, one of the genera I found shortly after arriving in Australia, only to find out Lindsay Jue Sue had already laid claim to it in his thesis. I named one species after my grad school friend, Russ Hobbs (*C. hobbsi*), and the second, *C. juesuei*, for obvious reasons.[46] My co-author, Vasyl Tkach, and another friend, Scott Snyder, later named a new species in the same genus, *Choanocotyle platti*, in my honor. My first patronym! I was deeply touched.

I found lots and lots of nematodes in the Australian turtles I necropsied but didn't want to describe them myself. *Camallanus* is a common nematode parasite of turtles worldwide and was a source of fascination early in my career. The anterior end of *Camallanus* is striking. Two symmetrical valves reminiscent of the logo of the Shell Oil Company surround the mouth. They are composed of tanned proteins and are golden brown. A quick perusal of the specimens I collected suggested two species, and both might be new. I started casting around for someone who might be interested in doing the work I wasn't keen to undertake.

I was chatting with a colleague at our annual meeting, and he suggested someone I was not familiar with — Mark Rigby. Mark earned a graduate degree with John Holmes (small world again) a decade after I left UA but never held an academic post. He obtained a position with a firm specializing in environmental assessments and bounced around in the field while moving up the corporate ladder. He has, however, maintained a near obsession with this particular group of nematodes. I found his contact information and e-mailed him straight away. Mark responded immediately and enthusiastically.

Despite Mark's lack of an academic position, and I would assume, a fully equipped laboratory, he does beautiful work. It took several years, but I finally received a manuscript ready to submit for publication. I was intrigued by the list of co-authors. I knew Ryan Hechinger, a graduate student of a long-time friend at the University of California — Santa Barbara. The other two, Jim Weaver and Reuben Sharma, were strangers, and I had no idea what contributions they made to the study. Reuben proved to be my gateway to a sabbatical, and my last collecting trip, in Malaysia (Chapter 22). The manuscript was accepted, and *Camallanus waelhreow* and *Camallanus nithoggi* entered the scientific bestiary.[52] The specific names have their origins in Old English (bloodthirsty) and Norse mythology (a serpent that gnaws at the root), respectively. Mark is much more creative in constructing names than I could ever be.

During my time in Australia, I necropsied 72 turtles and found 29 species of metazoan parasites. Metazoan means multicellular (worms). I don't look for protozoans. I had the data for an ecological examination of the Australian turtle/parasite system. That meant math. I learned early on mathematical work was not my strong suit, and I would need a collaborator with strong math skills. Derek Zelmer was that person. I met Derek in Costa Rica (Chapter 15) shortly after completing his PhD at Wake Forest University. Derek is a math whiz and loves large data sets. Once I identified and counted all the worms, Derek went to work. The resulting paper was well beyond my meager understanding of mathematical and statistical principles.[51] When you are in over your head, stop digging, and ask for help.

What about molecular biology? Remember, molecular biology was the main reason the Lilly Foundation so generously funded my year abroad.

I promised to learn the methods of this nascent technology and its application to trematode systematics and evolution. David Blair was to be my mentor, but his schedule precluded spending as much time teaching me the basics as he might have liked. A fair amount of my molecular education fell to his graduate students: Dani Tikel and Jess Morgan. Dani's area of interest was non-parasitological – dugongs (the Pacific version of the manatee), while Jess focused on trematodes. Both these young women were terrific mentors to an old guy; I was 44 at the time, trying to learn something new.

I worked hard under David, Dani, and Jess' tutelage. I spent over half my ten-month stint doing nothing but molecular work: extracting DNA, using PCR to amplify informative sections of the molecule, transferring the DNA segments into bacteria, and cloning them. I prepared and ran sequencing gels of the blood fluke specimens I brought from home and the Australian material from my trip to Brisbane. In the end, I got nothing useful, nothing elucidating the position of the Spirorchidae in the tree of life. The lesson I took away from all of this time and money? I didn't want to do molecular biology. I was a microscope guy.

Years later, Kathy and I visited Down House, Charles Darwin's home in rural England. As I carefully examined his study/laboratory, I realized Mr. Darwin could walk into my lab at Saint Mary's and understand what I was doing and how I was doing it with minimal explanation. I am confident he would marvel at the advances in microscope technology, but there would be nothing beyond his ken. His was the biology I loved. Spending hours collecting, staining, and studying specimens microscopically was where I belonged, not doing chemistry.

I learned the techniques of molecular biology but decided it wasn't for me. I often told students it is as important to discover what you don't like as it is to find your passion. I didn't share my insight with the folks at the Lilly Endowment. I did a bit of molecular biology upon my return to Saint Mary's and mailed the Lilly Endowment copies of all the papers published as the result of their largesse. Only one contained molecular work; and that was due to the efforts of my co-author. The Endowment never inquired about introducing molecular biology into Saint Mary's curriculum, and I certainly didn't offer. Don't ask, don't tell.

14. Student Research

A clever person solves a problem. A wise person avoids it. A stupid person makes it.

— Author Unknown

S mart versus clever. What is the difference? Is there a difference? I am certain psychologists have studied the question, but this is what I think based on nothing more than my intuition. Both are rooted in genetics and heritable. Smart is the ability to learn new things and apply the knowledge to related problems. Smart can be developed through education. Clever is the ability to use information in unique and surprising ways, thinking outside the box as it were. I don't think one develops clever the same way as smart. Maybe a little bit, but not as much. Asking someone to think outside the box who wasn't born with the facility is as cruel as asking me to solve a differential equation! I was born with modest amounts of smart and clever. Much of the student research I supervised relied more on the latter than the former.

The goal of any investigation is to increase our understanding of how the world works. To pull back the curtain and expose the breadth and depth of natural processes: ask a question to which the answer is unknown; construct a hypothesis delineating an either/or outcome; design experiments or observations allowing a choice between the two possibilities, nudging our understanding of the natural world forward. Sometimes a little, sometimes a lot.

Undergraduate research follows these dictums, although the scope most frequently falls on the lower end of the spectrum. While I participated in undergraduate research as a student at Hiram, I had no personal experience supervising students at this early stage in their education when I arrived at Saint Mary's. Working with undergraduates is vastly different from nurturing graduate students. Undergraduate students have heavy course loads, work requirements for student aid or need, and participation in the extra-curricular activities that make the college experience truly meaningful. Also, there is the problem of conflicts between the schedules of the student and professor. If I am busy on Mon-Wed-Fri and all the students' classes are on Tue-Thu, we have a problem. All these variables must be managed to provide sufficient time to produce a meaningful result and a paper approximating a manuscript suitable for submission to a scientific journal.

All biology majors at Saint Mary's completed a research project as a graduation requirement. Our department averaged between 20 and 40 seniors a year to be divided evenly among the faculty. If we had 24 seniors and eight faculty, each faculty member supervised three students during the year. One other thing — no cherry-picking. You got the first three students who walked through the door and said they wanted to work with you.

The system was in place when I arrived and remained virtually identical three decades later when I retired. Faculty circulated possible projects to the juniors before Thanksgiving break. After break, the students came to our offices and claimed projects on a first-come-first-served basis. Once we reached their allotted number, we directed stragglers to someone with open slots.

During second semester, all juniors took Biology 385 — Introduction to Research. Completing the course required compiling a bibliography on their research topic, writing a literature review, and finally writing and presenting a research proposal. The research proposal was a combination of the literature review plus the materials and methods required to generate the data forming the heart of the study. Students presented their projects to the faculty and class. The research advisor supervised the student's progress in each phase of the work. We edited, and they revised until they produced an acceptable paper.

First semester of the senior year was devoted to conducting the experiments or observations, data collection and analysis, and writing the first draft of the paper. During Christmas break, we reviewed their efforts and provided editorial guidance, followed by multiple revisions. Once the advisor was satisfied, two different faculty assessed the writing and made suggestions for clarification and a final rewrite. As a capstone experience, the student presented her research to the entire department as both oral and poster presentations. Students earned three credits toward their major and graduation requirements. And the faculty? The faculty received nothing: no teaching credit, no release time, and no money to fund the research. It didn't matter. Every year two to five new students appeared in your office expecting a research experience, and every year they all got one.

Most students had little or no familiarity with the process, and many had no background in the discipline they were investigating. They were blank slates. Students ran the gamut of intelligence, creativity, and commitment. You might get next year's valedictorian or someone barely scraping by. It didn't matter. Our job was to ensure every student completed a project to the minimum departmental standard.

I quickly discovered most biology majors at Saint Mary's aspired to careers in some medical field: physician, dentist, physical therapist, and the like. They wanted a research experience related to those goals. Most found my beloved taxonomy a snooze. Well over half of my students studied aspects of the life cycle of the intestinal trematode *Echinostoma caproni*. However, some worked on ecological studies of land snails, the migration of tapeworms in the rat intestine, an unusual leech-host relationship, a couple of taxonomic studies, the presence of nematode eggs in public parks, and other miscellaneous subjects. During my 28 years at Saint Mary's, I supervised 76 projects, with 15 students contributing to 12 papers published in national and international journals.

For the first 20 years, I successfully shepherded students through the process with a diversity of projects; some produced interesting results but were hampered by small sample sizes, precluding any possibility of publication. Many more failed to falsify the null hypothesis and didn't tell us anything interesting. Three students produced publishable work, although generally several years after they graduated and significant reworking by me. All of our students did a remarkable amount of work and produced papers and presentations they could look back on with pride. Those resulting in publication weren't different in commitment or effort; they were lucky in their choice of projects. Here is a brief overview of the first three resulting in publications.

Looking for Worm Eggs in Dirt[21]

Karen Ludlam, my first research student, examined the soil of three parks in St. Joseph, MI, for contamination by the eggs of *Toxocara canis*, a large roundworm parasitizing dogs. Infection with the larvae of *T. canis* in humans may cause a disease called visceral larva migrans (VLM), which occurs predominantly in young children. VLM was one of the reasons for the passage of pooper-scooper laws. Children acquire the infection by eating dirt contaminated by dog feces. The eggs can survive in the soil long after any sign of fecal contamination disappeared. Eggs ingested with the soil hatch in the child's intestine, migrate through the gut wall, and wander around the body. Damage to the child depends on the number of eggs ingested: small numbers are typically not harmful, but large numbers may cause death.

We chose three parks based on observed levels of maintenance and whether dogs were permitted or not. Karen collected soil samples from each location in areas where children were likely to play. She processed the soil to isolate the eggs and compared the numbers between the three sites. Despite differences in the area's level of care based on emptying of trash cans, the cleanliness of restrooms, and whether dogs were permitted in the park, Karen found *Toxocara* eggs at similar levels in all three locations. Karen went on to earn a PhD in agronomy and is a practicing scientist.

Leeches and Salamanders[28]

In the spring of 1990, Dave Sever, our anatomist, walked by my office and asked, "What do you know about leeches?" I replied, "Nothing." Dave studied tiger salamanders (*Ambystoma tigrinum*) from a temporary pond at Holy Cross College, a half-mile south of Saint Mary's. Adult salamanders are terrestrial burrowers but congregated in the pond to mate and lay their eggs in early spring, hoping the young salamanders would be ready to take up their burrowing lifestyle before the pond disappeared between late May and mid-June. Dave collected specimens every year for 15 years and never saw a leech. That spring, they swarmed the small pond.

Leeches are not easy to identify by non-specialists. I got Roy Sawyers's three-volume treatise on the Hirudinea from the library and got to work. The leech Dave found turned out to be *Helobdella stagnalis*, perhaps the easiest leech in the world to identify, although recent molecular work suggests that may no longer be true. This leech is a predator on snails and other invertebrates, not a blood feeder; however, many salamanders had numerous leeches attached to their bodies. Dave and I thought exploring this association would be a good student project. *Helobdella stagnalis* had never been reported parasitizing a salamander.

Veronica Gonzalez, a student in Dave's electron microscopy class, had the requisite histological training and expressed interest in the project. One of the few times I violated the cherry-picking rule. The following spring, we collected leeches from salamanders and prepared them for identification and histological examination. Veronica embedded specimens in paraffin and cut the leeches into 10 μm sections. She examined the intestine for amphibian red blood cells, a sure sign the leeches were feeding on the host. She didn't find any, which was puzzling. The leeches were firmly attached to the skin, clustered near lymph sacs; however, lymph does not contain red blood cells. We could not make the case the leeches were feeding. The following year the parasites were gone and never reappeared during Dave's tenure at Saint Mary's. Where they came from and why they disappeared remains a mystery.

Movement of Tapeworms in Dead Rats (or Post-mortem Migration)[31]

Parasite ecologists study the distribution of parasites in the host, especially in the small intestine. In many cases, we can't (for logistical reasons) begin a necropsy until the host has been dead for several hours, raising the issue of post-mortem migration. Do parasites move from their preferred locations in the host when they "realize" the host is dead? The parasite equivalent of deserting a sinking ship. I thought this might make an interesting student project using the rat tapeworm, *Hymenolepis diminuta*. Susan Villanueva volunteered.

The eggs of *H. diminuta* develop into larvae called cysticercoids when ingested by flour beetles, *Tribolium spp.* The cysticercoids produce adult tapeworms when a rat consumes an infected beetle. I bought infected flour beetles from a biological supply company and rats from a different outfit supplying rodents for biological research. We infected 27 rats with five cysticercoids each and waited 28 days for the establishment of mature infections.

While we could have dissected the intestines when we removed them, we opted for a procedure I learned from John Holmes to make the process more convenient: fast-freezing the intestine. If you place dry ice (frozen carbon dioxide) in 70% ethanol, the alcohol temperature drops to about −70°F. Place a long, thin intestine into the mixture, and it freezes almost instantaneously. The frozen gut can be placed in a labeled ziplock bag and stored in a freezer for later examination. Gut length is maintained, and the worms are frozen in place: ideal for our study.

We killed one group of three animals at 10:00 am. They served as the starting point for both the experimental and control groups. All twelve rats in the experimental group were killed simultaneously and allowed to remain at room temperature until necropsy. Susan dissected groups of three animals 30, 60, 120, and 240 minutes after death and froze their intestines. We divided the remaining twelve rats into groups of three. One group was killed and processed at the same intervals as our experimental animals. So, we had three animals to tell us where the worms were at 10:00 am under normal conditions. The control group would tell us where the worms were at 10:30 am, 11:00 am, 12:00 pm and 2:00 pm in living rats. The experimental groups would tell us if the worms moved in dead animals compared to where they were at 10:00 am and compared to where they should be at the remaining times if the host hadn't died.

Susan found her worms did move anteriorly in the intestine between one and two hours after the rat died, suggesting necropsies should be performed as soon after the host's death as possible to obtain the most accurate assessment of parasite location for ecological studies. Only one other study, to my knowledge, examined this phenomenon. Bernie Fried, of Layfayette University and undisputed king of *E. caproni* research, conducted a similar experiment with the trematode, *Echinostoma caproni*. He failed to detect any change in the worms' location 36 hours after the host's death. Unfortunately, Bernie used a coarse-grained approach (dividing the gut into five equal sections) to determine parasite location that might mask the parasites' more subtle movement. Susan went on to a successful career as a compliance officer for a large pharmaceutical firm.

15. Costa Rica

When you're in jail, a good friend will be trying to bail you out. A best friend will be in the cell next to you saying, 'Damn, that was fun.'
— Groucho Marx

Dan Brooks and I met at the 53rd Annual Meeting of the American Society of Parasitologists in Chicago in 1978. We were newly minted PhDs and bonded instantly. Dan was a prolific author turning out manuscripts and books at a prodigious pace. He never thought small. Cataloging Life on Earth became a "thing" in the 1990s, and Dan wanted in. Dan never lacked for high-powered collaborators and hooked up with Dan Janzen, University of Pennsylvania, to work at the Guanacaste Wildlife Area in northwestern Costa Rica.

Janzen was an enigmatic figure in the world of ecology, at least to me. Janzen didn't think small either. Enamored of tropical ecosystems, he worked to acquire land in Costa Rica to establish and expand a refuge where like-minded people could catalog all life forms in a small corner of the world. He arranged to spend half the year teaching and the other half working on projects in Central America. Brooks signed on to catalog the parasites of vertebrates in Guanacaste. Dan estimated the 900+ species of vertebrates harbored over 11,000 species of parasites, a project that could take him to retirement.

Dan turned his grant money into salaries to train and employ locals as parataxonomists. As paramedics provide medical treatment before the patient reaches the hospital, parataxonomists conduct necropsies and make preliminary identifications of the worms they found. Appropriate specialists received the specimens for identification and then describe any new taxa present.

Brooks contacted me in mid-1997 with the offer to join him and collect turtle parasites in Guanacaste the following summer. All I had to do was finance my trip. I have a natural aversion to grants, but this was too good to pass up. The lure was the possibility of being the first person to find TBFs in Central American turtles, and I would get to hang out with parasitologists for two weeks. Saint Mary's offered small, competitive grants for teaching and research. I decided to bite the bullet and submit an application. Fortunately, the requirements for this in-house money were not terribly rigorous. I wrote a short overview of the "inventory" goals and my small part in the enterprise. I figured the cost of travel plus room and board at about $1,250. I wasn't surprised when I received the award. However, I was surprised when one of my friends on the Grants Committee shared that another member, also a friend, argued against my application because I would kill turtles. I take no pleasure in killing animals, but if we want to learn about the diversity of life on Earth, that is part of the process. Fortunately, none of the other members concurred.

In early June 1998, I gathered the supplies I needed for two weeks of dissecting turtles and preserving the parasites I knew I would find for transport back home. I packed snorkeling gear if the traditional hoop nets failed to provide the specimens I hoped to examine. Beyond those items, all I needed were clothes and some books to read during any downtime. The lab had electricity but no electronic entertainment.

The headquarters of the conservation area contained a dormitory, dining area, and miscellaneous structures. And, of course, the "jaula" or cage, where we conducted necropsies. The jaula consisted of a dirt floor and two enclosed walls, and the remaining two covered with chicken wire — hence the nickname. There was a sheet metal roof in case of rain. We had a sink with running water and overhead lights, which turned out to be a blessing and a curse. The folks at Guanacaste built the facility for Brooks for the express purpose of parasitological examination of whatever he caught. It met the bare minimum standard for what we needed to do. Lab benches with electrical outlets for the dissecting microscopes, tables for conducting dissections, and we didn't have to worry if blood or anything else hit the ground.

Each year Dan invited a different gaggle of friends and acquaintances to join him in the quest to catalog the parasites infecting the local vertebrate fauna. In 1998, the group included Anindo Choudhury, a postdoctoral fellow of Dan's specializing in fish, Derek Zelmer, who recently completed his PhD dissecting amphibians, and me doing turtles. The only person common to all of us was Dan. We all got along famously. I wrote letters of recommendation for Anindo during his job search. He was hired by St. Norbert's College, Wisconsin, and has had a stellar academic career. Derek and I developed a close working relationship spanning a decade and a half and co-authored over a dozen papers. On the other hand, while Dan and I keep in touch sporadically, we went our separate ways as people frequently do.

Once the team assembled, the first order of business was to collect animals to dissect. Different animals required different methods of attack. Fish were easy if you had an electro-shocker, a battery-operated device sending an electrical charge into the water, stunning the fish. The comatose fish float to the surface and were scooped up with a net. Frogs were trickier. Some folks use nets, but frogs are wary and frequently escape into the water. Derek preferred catching them by hand. Derek stalked his target, moving slowly, and at the last moment, quickly grabbed the unsuspecting animal. The frog probably did suspect something was going on but figured it would avoid detection if it remained motionless. The frog was wrong. Derek rarely missed, mimicking *T. rex* snatching the lawyer off the toilet in *Jurassic Park*.

The first two days did not yield any turtles, and I was frustrated. I went for a walk to scout the nearby area for potentially more fruitful collecting sites. As I returned to the lab, I saw a turtle scampering across the lawn toward the underbrush. I took off running and grabbed it with a genuine sense of delight. Then I heard the laughter. Dan, Anindo, and Derek emerged from behind a building shrieking hysterically. They caught the turtle in one of the small, ephemeral ponds (really mud holes) dotting the forest floor near the lab. They saw me coming and released the little bugger, hoping I would do exactly what I did. I was the day's entertainment.

We captured another half a dozen in similar spots within easy walking distance of the lab.

Our workdays varied depending on who needed what. Collecting fish was a daytime activity, as was turtle hunting. Amphibians were easier to catch at night. Males typically call for mates beginning at dusk, and their eyes are easy to spot with a flashlight in the dark. Collecting was not without its hazards. During the day, we were on high alert for snakes. There were highly venomous species in the area. We were fortunate not to have encountered any as none of us wore snake leggings to protect against bites. At night, leopards were on the prowl. We didn't see any, but it didn't mean they didn't see us.

One morning I accompanied the fish folks to a nearby stream for one of their electro-fishing expeditions. While they worked the rapids knocking fish senseless with electricity, I went upstream to try my luck snorkeling for turtles. Hey, it worked in Australia. The stream (the name long forgotten) was about 20 feet wide and 3–4 feet deep. I stripped down to my swimsuit and donned mask, fins, and snorkel. Underwater visibility was limited as I started in a large deep pool and continued as it narrowed, feeling under the overhanging bank for turtles. I worked my way about 100 yards upstream, crossed to the opposite bank, and repeated the process heading back toward my starting point. About 20 yards from the rapids where the fish folks were working, a large boulder protruded into the pool where I began. As I followed the rock's contour, I was looking down (for turtles, naturally). As the boulder curved back toward the shore, I raised my head and saw an alien-looking eye. In the next nanosecond, I realized the eye was part of a caiman (an alligator relative), and we were about to have an intimate encounter.

The power generated by a burst of adrenalin is nothing short of amazing. I exited the water like a Polaris missile. Straight up! I hit dry ground and started climbing — no small feat wearing flippers. Fortunately, the caiman felt the same sense of alarm. When I went up, he (or she) headed for deeper water. I estimated the length of the critter at about six feet, large for *Caiman crocodylus*, the spotted caiman, common in Costa Rican waters. Like many anglers, perhaps I exaggerated the caiman's size. However big it was, the animal was dangerous. I escaped with a few scratches from scrambling up the bank in my swimsuit — and a great story. However, it was my last foray snorkeling for turtles in Central America.

Guanacaste's living quarters were spartan, but we weren't there for the turndown service and a chocolate on the pillow. The dorms were concrete block construction, with bunk beds (two per room) and not much else. The washroom consisted of communal showers, a line of sinks with mirrors for shaving, and commodes. I was warned upon arrival to shake out my shoes every morning and inspect the underside of the toilet seats before taking care of business to avoid scorpion stings. A warning I heeded religiously.

We conducted most necropsies at night as the majority of collecting took place during the day. We typically spent our time after dinner until about 10 pm in the jaula. It was the dry season, so rain wasn't a concern, but insects were. The overhead lights consisted of a metal shade with a single, incandescent bulb. Insects were attracted to light, so the staccato of bugs banging into the light shades provided a rhythmic tempo to our activities; some flew off or fell into our workspace. Most were harmless.

Blister beetles were in a category all their own. Those small, nondescript insects secrete a chemical, cantharidin, a defense against predation, and a potent skin irritant when crushed. It seemed when they flew into the lights, they gave up and dropped straight down. The natural response when something lands on you is to smack it. However, if you squash a blister beetle, it releases more irritants producing a painful chemical burn; not reacting required exceptional control.

We only caught two species of turtles: *Kinosternon leucostomum* and *Rhinoclemmys pulcherrima*. Species of *Kinosternon*, commonly referred to as mud turtles, are classified in the family Emydidae and restricted to the western hemisphere. Species of *Rhinoclemmys*, or wood turtles, are classified in the family Geoemydidae. Geoemydids are common in India and Asia. *Rhinoclemmys* spp. are the only geoemydids found in the western hemisphere.

Neither species harbored large numbers of worms in the small intestine and associated organs. The mud turtles yielded TBFs, *Hapalorhynchus albertoi*, first described from a related turtle in Mexico in 1978, but I was the first to find them in Central America. I came up empty for TBFs in wood turtles. I received quite a shock when I opened the large intestine of the first wood turtle. It contained a writhing mass of nematodes. Thousands of nematodes clogged the posterior portion of the gut. I collected, fixed, and stored as many as possible. Every wood turtle I dissected revealed the same state of affairs, more nematodes than would fit in the small vials I brought with me. Cursory examination suggested there were probably three species present in every animal.

Monogenetic trematodes, *Neopolystoma* sp., occurred in the conjunctival sac of both species of turtles. I was excited as polystomes were unknown from this location in turtles east of the International Date Line. I was sure I had at least one new species, possibly two. I found a few other species of nematodes and trematodes, but they were familiar and almost certainly represented species already reported in the literature. I had more than 100 vials of specimens to examine when I got back to Saint Mary's.

Finally, it was time to drive back to San José and wait for my flight to depart. I wouldn't trust my worms to the vagaries of checked baggage, so I had the box of vials in my backpack as I passed through security to the boarding area. Dan arranged for the permits I needed to remove biological specimens from the country. However, the local officials regarded me with a mixture of concern and curiosity when they opened my backpack and examined the vials filled with preservative and parasites. After 9/11, I would never have been allowed to board a plane with them.

The nematodes were fascinating. While I didn't want to do the complete workup myself, I needed to get a handle on what I had and find a specialist willing to describe them. I prepared a few worms I deemed different from each other (I thought three species) and worked out the family to which they belonged. Species of the superfamily Cosmoceroidea (the suffix -iodea indicates the superfamily designation in zoological circles) and the family Atractidae (the same for the suffix -idae for families) are frequent denizens of the large intestine of various vertebrates. I took a stab at identifying them to the generic level with varying levels of confidence. I had a good idea of my first choice for a collaborator, Lynda Gibbons, Veterinary College, University of London. I never met Lynda, but we corresponded over the years, so it wouldn't be a cold call. Lynda co-edited keys to the genera of nematode parasites of vertebrates in the 1970s and '80s. She also did a great deal of work on atractid nematodes. My philosophy has always been when you ask for a favor, the worst that can happen is the answer is "No." Rejection is not the worst thing in the world.

I sent the request, and Lynda accepted. She warned me she had other projects in progress, and it might take some time to get to mine — fine by me. I doubted anyone else was collecting turtle parasites in Costa Rica. In 2006 a paper appeared in the British *Journal of Helminthology* entitled "*Rhinoclemmysnema* n.g. and three new species of nematodes of the family Atractidae (Cosmocercoidea), with notes on the helminth fauna of *Rhinoclemmys pulcherrima* (Testudines: Bataguridae) in Costa Rica."[48] Lynda did her usual excellent job. I couldn't have picked a better person to whom to entrust those intriguing worms.

The massive number of worms found in every animal I examined raised several questions, none of which I could answer definitively. The method of host-to-host transfer for these worms is unknown, but related species infect the host directly by ingestion of eggs or larvae. Since every turtle harbored massive numbers of worms, the infective stages must occur in significant numbers where the turtles feed. The huge populations also might be the result of autoinfection. The adults of similar species release larvae in the gut that mature without ever exiting the host. While the sample size was small (I only necropsied seven mud turtles), every individual was heavily infected, suggesting the worms might benefit the turtle. If they assisted in breaking down the plant material making up the bulk of the mud turtle diet, it might represent a mutualistic (both partners benefit) relationship. This guess was a bit of a stretch, but who knows? If my musing leads someone to test the hypothesis, it was worth the ink and paper.

Work on *Neopolystoma* from the conjunctival sac proceeded a bit more quickly. I had been studying the parasites of painted turtles from the Upper Peninsula of Michigan and found similar worms in their eyes. I had a nearly complete manuscript describing a new species, *Neopolystoma elizabethae*, so the literature was fresh in my mind.[40] All I had to do was prepare and measure the specimens and do the required drawings. I decided to christen the worm from south of the border, *Neopolystoma fentoni* (Figure 7).[41] These two were the first species of *Neopolystoma* found in the conjunctival sac of turtles from the western hemisphere.

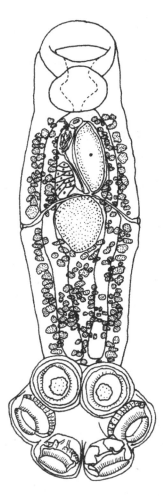

Figure 7. *Neopolystoma fentoni* from: Platt, T.R. 2000. *Neopolystoma fentoni* n. sp. (Monogenea: Polystomatidae) a parasite of the conjunctival sac of freshwater turtles in Costa Rica. *Memórias do Instituto Oswaldo Cruz* 95: 833–837. (Reprinted with permission.)

The names I chose for the new species of monogenetic trematodes warrant comment. My wife's maiden name is Fenton. I did not name a parasite after my wife; however, she wouldn't have objected. We had been married for over a quarter-century, and she knew naming a new species after someone was an honor. I named it for her father, Bill Fenton — a kind and delightful man. Bill was in his late 80s and in failing health. I thought he would appreciate having a new species named for him. Dad Fenton was thrilled. I was glad I could do something to provide a small measure of immortality for him before he passed away.

Neopolystoma elizabethae was a different matter. Our church, the First Unitarian Church of South Bend, holds an annual service auction to raise money for operations and outreach. Most people donate dinners and services for other members to purchase. I decided to offer the naming rights to a new species of parasite. I would name the worm according to the wishes of the highest bidder. I didn't invent this scheme. Scientists have named new species in honor of their patrons for centuries. More recently, scientific societies offered to name new taxa after their benefactors to raise funds for research. Since I can't cook and have no other saleable skills, I thought I could raise a few bucks for the church through my professional activity.

The bidding was spirited, and I was pleased until my wife informed me one of the bidders wanted to dis her ex-husband by naming a parasite after him. The International Code of Zoological Nomenclature prohibits the formation of names considered vulgar or derogatory.

That didn't stop two entomologists from naming a pair of slime mold beetles after George W. Bush and Dick Cheney!

Even if the Code didn't have that prohibition, I view naming new taxa after a person as an honor, and I wouldn't violate my personal beliefs for any amount of money. As the bidding passed the $150 mark, I started to get nervous. Fortunately, the woman scorned and dropped out. The winner, Elizabeth Scarborough, was a dear friend, former Dean of Arts and Letters at Indiana University South Bend, and an avid amateur naturalist. I was delighted to name *Neopolystoma elizabethae* in her honor. In addition to the name, she received a framed photograph of my drawing of the worm to hang in her townhouse and a copy of the article. She was thrilled, and I was relieved!

Brooks' parasite inventory gained some notoriety in Carl Zimmer's magnificent paean to parasitology, *Parasite Rex*, published in 2000 and is still in print. Zimmer visited Guanacaste in 1997, the year before my participation, but, alas, the project ended before the book was published. The inventory collapsed due to lack of funding, and Dan's dream of cataloging all the species of parasites evaporated as well. Undeterred, when climate change and emerging infectious diseases took center stage, Dan took up the challenge.

16. Are You Smarter than a Trematode? I — Cercariae

If you do an experiment and it gives you what you did not expect, it is a discovery.

— Martin Chalfie

I n the late 2000s, student research began yielding publishable results. Over my last decade in the department, 12 of my students contributed data resulting in nine publications. What changed? First, I think I got better at identifying interesting questions using *E. caproni*, and the addition of Derek Zelmer as my math and statistics guru was a significant factor. As I indicated on more than one occasion, I am a mathematical ignoramus. My grades in high school math were adequate; I earned B's. I took one calculus course in college and received a C. I can reason biologically, but equations don't make any sense to me. I can't look at a formula and intuit what it represents. I can do basic statistics and explain the rationale to students. Anything more sophisticated, and I am lost. Derek is a whiz. More importantly, Derek is a parasitologist. Having a statistician who intimately understood the systems my students studied was revelatory.

I met Derek in Costa Rica, as described in the last chapter. We got on well but weren't close. A few years after our initial encounter, a student of mine, Elizabeth Warburton, informed me near the end of senior year of her desire to pursue a Master's degree in Parasitology. I didn't see much hope for her getting accepted for the fall semester as most schools' application deadline had passed. I thought about Derek. I knew he took a position at Emporia State University, a small state school in Kansas. I guessed they weren't an academic powerhouse and might take a student late in the game. I knew Derek was terrific, and the advisor often means more than the institution for a solid graduate education. I called him. We spent a few minutes catching up; then, I posed the question of Elizabeth. He shared he was on the graduate committee, and they were still reviewing applications. He encouraged Elizabeth to apply. She did and was accepted. She completed her Master's with Derek and went on to earn a PhD elsewhere.

At the same time, I was collaborating with another old acquaintance who had fallen off the parasitological grid. I enlisted him to undertake an ecological assessment of snapping turtle parasites from the University of Richmond. After a few years of inactivity, I had to cut him loose. I asked Derek if he was interested in taking a crack at it. He jumped at the chance, and we had a paper in print a year later.[55] I had a math guy!

The focus of student research in the last decade of my career was experimental investigations of the life cycle of *Echinostoma caproni*. We designed and conducted the experiments, and once we completed the data collection, Derek performed the statistical analysis.

Adult *E. caproni* reside in the posterior portion of the small intestine of mice. Eggs are passed in the feces and develop to form miracidia, a free-swimming stage that infects the freshwater snail, *Biomphalaria glabrata*, and reproduce asexually. Asexual reproduction results in thousands of free-swimming cercariae, which escape from the snail and must find a second intermediate host. Fortunately, *B. glabrata* is also a wonderful second intermediate host. The free-swimming cercariae penetrate the snail and encyst forming metacercariae. If a mouse ingests the snail, each metacercaria has the potential to produce an adult worm.

These projects can be divided into either cercarial studies or examining the activity of adult worms in the small intestine of mice. I'll start with the cercariae and save the adults for later.

Larval Trematodes — Light and Gravity[54]

The real breakthrough in studying the life cycle of *E. caproni* came in 2008. I had an idea to examine the effect of environmental factors on the transmission of the cercariae of *E. caproni* from the first to the second intermediate host. Knowing how cercariae respond to light/darkness, gravity, temperature, currents, and other environmental cues is critical to understanding how trematode life cycles function and offer clues to possibly reduce or stop transmission. Unfortunately, most studies of this nature only examine behavior. Do the cercariae respond positively or negatively to light? Do they swim up or down? What water temperature is preferred, and so on. They rarely assessed the actual infection of the next host. Understanding behavior is helpful, but knowing how behavior affects transmission is critical.

Echinostoma caproni uses the snail, *Biomphalaria glabrata,* as both the first and second intermediate hosts. I needed to find a way to separate infected snails releasing cercariae (shedding snails, or 1st host) from uninfected snails targeted for infection (sentinel snails, or 2nd host) by the free-swimming larval stage. I had an idea for a transmission chamber, but I needed a transparent tube, either glass or plastic, as the foundation for the contraption I had in mind.

While perusing the catalog of an aquarium supply company, I found something with promise — clear PVC pipe. The size seemed right, and I could easily cut it into suitable lengths. As a bonus, there were also slip couplers to connect the pieces and "T"-connectors to make more elaborate choice chambers. There was one thing I had to know — the internal diameter. I envisioned using round tissue cassettes to hold the shedding and sentinel (uninfected) snails in precise locations. These are small, plastic devices used to process tissue for histological preparations, and a former faculty member left thousands of them when he took a position elsewhere. Each cassette consisted of two pieces about 1 cm thick when snapped together, with crosshatched openings to allow the flow of chemicals as tissue samples move through a series of solutions. The holes would also permit cercariae to swim in and infect a snail, but the snail could not escape. The cassettes were 38 mm in diameter, so the PVC pipe had to be 39–40 mm internally. I called the supplier, asked the question, and held my breath. The response? 39 mm! We were in business.

We used silicone sealer to glue a slip coupler onto a Petri dish as a base and inserted a piece of tubing about 12 inches long to create a vertical transmission chamber (Figure 8). We used a thin glass rod to skewer three tissue capsules positioned near the top, middle, and bottom of the water column.

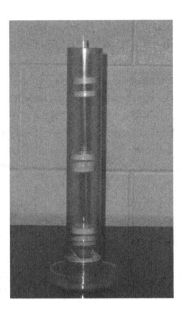

Figure 8.   Choice chamber for gravity from: Platt, T.R., L. Burnside, L. Bush. 2009. The role of light and gravity in the experimental transmission of *Echinostoma caproni* (Digenea: Echinostomatidae) cercariae to the second intermediate, *Biomphalaria glabrata* (Gastropoda: Pulmonata). *Journal of Parasitology* 95: 512–516. (Reprinted with permission.)

The top and bottom capsules would hold the uninfected, sentinel snails, while the middle one would contain the snails shedding cercariae. Fill the whole thing with water, and we would be able to determine if the cercariae preferentially infected snails at the top or bottom of the water column. The only problem was the tube and slip coupler did not form a watertight seal. I couldn't use aquarium cement because I wanted to be able to disassemble the pieces. I thought about it, and the solution came in a flash, the second greatest invention in human history (the remote control is first) — Teflon™ tape. That stuff allowed almost anyone to do basic plumbing, and a single wrap around the bottom of the vertical tube sealed the system, no leaks, and easily disassembled. Lindsay Burnside and Elizabeth Bush expressed interest in this project and were excited about the idea.

The concept was simple and would answer an intriguing question. Does behavior predict transmission? Previous studies indicated *E. caproni* cercariae were positively phototactic and negatively geotactic. This technical jargon means the cercariae swim to the top of the water column toward light and away from the Earth's center. If we placed shedding snails in the center of the transmission chamber and sentinel snails at the top and bottom (contained in tissue capsules to ensure they stayed in place), the only things moving were the cercariae. Wait for a few hours to give the cercariae a chance to disperse in the water column, remove the snails and count the number of metacercariae in snails at the top and bottom, and we would have an answer for gravity. Since each student needed a separate experiment, we devised a similar system to test the response to light. We glued the tops of small glass Petri dishes on slip couplers to form a sealed chamber, wrapped half of the chamber in aluminum foil to exclude light, and laid the chamber parallel to the lab table.

At the end of each experiment, the student extracted the glass rod with the tissue capsules attached, placed the sentinel snails from the top and bottom (or light and dark) in separate beakers, and let them sit overnight to allow the cercariae to encyst. The next day the students removed the shell from each snail, placed the body on a microscope slide, and smushed it with a second slide. The students examined the flattened snail using a dissecting microscope and counted the number of metacercariae. I had enough tubing to construct four chambers for each experiment to run multiple trials and collect enough data for statistical analysis. Both experiments shared the following attributes: a clear hypothesis, easily obtained data, and the students could complete multiple replicates in a few weeks.

Our results supported the behavioral studies for gravity but not for light. More metacercariae infected snails at the top of the water column suggesting negative geotaxis. However, snails in the dark were more heavily infected, indicating a negative phototaxis. Interestingly, we found a significant minority of metacercariae in snails at the bottom of the water column and in the light.

Aquatic snails occupy diverse locations in aquatic ecosystems. They move in response to changes in temperature, time of day, seasonally, and attempted predation. If cercariae were monolithic in their search for a suitable intermediate host, they might miss opportunities to continue the life cycle elsewhere. While exploiting the environmental cues that proved advantageous to their cercarial ancestors was an effective strategy, a more extensive exploration of the ecospace and taking advantage of opportunities in less favorable locations would expand the pool of infected snails available to the definitive host, the animal eating the infected snail and harboring the adult worms. The aversion to light was, in the words of Issac Asimov, funny. These experiments provided a better understanding of the transmission of *E. caproni* and opened the door to more questions and more student projects.

Elizabeth Bush is an optometrist, and Lindsay Burnside was a laboratory technician at Notre Dame.

Larval Trematodes — Light versus Gravity: Part II[56]

Cercariae go up in the presence of overhead light. Which environmental cue is more important — light or gravity? Separating those environmental constants would require either access to the International Space Station or some serious thought. There isn't much you can do to alter gravity, but light is another matter. The configuration of our darkroom provided a possibility. The counter design included a movable shelf allowing a light source to be placed either above or below the transmission chambers. I thought running the same experiment in the dark or light coming from above or below the chambers might yield interesting results. I added the project to the list available to students starting their senior research projects in 2008.

Hali Greenlee appeared in my office after Thanksgiving and expressed interest. Hali was an unusual student for Saint Mary's. She was a little older, lived off-campus, was married, and had children. Hali came from a decidedly blue-collar background. Her husband worked as a carpenter, and I think she was the first person in her family to attend college. Hali was also whip-smart. Since I only taught Parasitology in odd-numbered years, Hali couldn't take the course until the following year, but she earned the highest grade in the class when she did.

We reviewed the results from previous experiments and discussed the rationale and design of the research I envisioned. As I explained the setup in the darkroom, we both realized light was going to be a problem. We wanted light to come directly from the top or bottom of the tubes with no extraneous light from the sides. The solution was cheap and straightforward, both critical factors when money is tight. Hali made cylinders of black construction paper that slid easily over the PVC tubing. We conducted the trials in a photographic darkroom with light entering from the top of the tube, the bottom of the tube, or in complete darkness.

The results supported our earlier findings. Transmission was higher in sentinel snails near the surface with the light at the top. With the light at the bottom, or with no light at all, the numbers were significantly smaller but not different from each other. So cercariae respond to a contradictory environmental cue (light coming from the wrong place — the bottom) as if there was no light at all. The numbers for sentinel snails at the bottom of the transmission chamber were precisely the reverse: few infections when the light was at the top, but significantly more when the light was in the wrong place (the bottom) or absent. It appeared light and gravity provided an additive effect drawing cercariae to the surface where snails were more likely to congregate. When light came from the wrong place or was absent, more cercariae seem to hang around the bottom, raising transmission levels. Why? I had no idea, but snails might move down in the water column during the evening, and the chance of transmission was higher in deeper water as the sun set. Again, the results of this experiment answered some questions and raised more.

Larval Trematodes: A Behavior Study — Why Not?[60]

My interest in studying the transmission of *E. caproni* cercariae stemmed from my dissatisfaction with behavior studies. Behavior studies can be quite valuable. The question that intrigued me was how behavior (i.e., response to environmental cues) affected transmission (i.e., infection of the next host). However, I thought one behavior study would make a good student project — simple and cheap. Take a small Petri dish with half (top and bottom) covered with electrical tape to exclude light from one side while the other half was illuminated, add 20 cercariae and see which half they prefer over time. This experiment appeared in a paper in the late 1990s using a related species of trematode, *Echinoparyphium recurvatum*. It seemed straightforward, so I put it on the list.

Anna Kammrath expressed interest and reviewed the literature. We had everything in place and started the observations; however, there were two problems. The Petri dish was too large to view in a single field on the dissecting microscope, and the cercariae were exceedingly difficult to count accurately. When Anna attempted to count the cercariae, she couldn't find all of them fast enough to know where they were at the end of the two-hour test period. Neither could I. I could not, for the life of me, understand how the author of the original study accurately counted the little buggers. We abandoned the project, and Anna moved on to something else.

I still thought this was a good project; however, I had to devise a method for accurately observing the cercariae. About a year later, I had a small "eureka" moment and grabbed the most recent edition of the Carolina Biological Supply catalog. The reason for my excitement was the recollection that Carolina sold a small plastic slide with a built-in chamber that might work. I quickly found the slide section, and there it was — the Carolina Deep-Well Slide®. The slide consisted of a square plastic base with a raised circular ring in the center, with a lid forming a closed chamber holding about 1 mL of liquid. Put a bit of electrical tape to cover half the well on the bottom of the base, do the same to the lid, and you have a light/dark choice chamber a student could observe in its entirety with a dissecting microscope. Best of all, they cost less than two bucks each. The project, in modified form, went back on the list.

First through the door was Rose Dowd. Rose was a free-spirit with a wardrobe better suited to an early Cyndi Lauper performance than a science lab. Rose was not the strongest student. For some reason I no longer recall, Rose never took Parasitology, which was probably a blessing in disguise. Regardless, she had to do a research project, and Rose chose me to guide her.

We talked about problems with the original experiment and made some significant improvements (in my mind) to the design. Rose would follow one cercaria at a time, much easier than trying to keep track of 20. Instead of a single time interval, she evaluated four. Rose also kept track of how many times each cercaria crossed from light to dark and back again. Rose had a stopwatch, a mechanical counter, and a kitchen timer. The cost of the entire experiment was about $20.

We placed an infected snail in a small volume of water and waited for cercariae to emerge. As soon as one came out, we captured it with a pipet and placed it in the well of the slide, and secured the lid. Rose put the slide on the stage of the microscope and located the cercaria. Once she found it, Rose started the kitchen timer for a countdown of five minutes and started the stopwatch. When the cercaria crossed into the dark, Rose paused the stopwatch and clicked the handheld counter. When it emerged from the dark, she clicked the counter and restarted the stopwatch. This continued until the kitchen timer announced the end of five minutes. She had the amount of time the cercariae spent in the light (and dark) and the number of times it traversed the light/dark boundary in five minutes. Rose set the first slide aside and constructed a new one. Rose repeated the process until she had five slides going at one time. She had a short break before repeating the process for each cercaria at 1 hour, 2 hours, and 4 hours post-emergence. At the end of the project, Rose had the time 20 cercariae spent in the light (and dark), and the number of crossings for each one at four-time intervals.

Our study supported the conclusion of the previous studies: cercariae became negatively phototactic (i.e., they preferred the dark) as they aged. However, with the additional data points plus the information on time, Rose added some nuance to our understanding of cercarial behavior. When cercariae first emerge from the snail, they don't seem to care about light and dark at all. They were highly active, moving between light and dark with reckless abandon. They spent about the same amount of time in the light and dark halves of the chamber. After one and two hours of swimming, those values dropped by half. After four hours, the cercariae spent almost all their time in the dark and rarely crossed back into the light. What might this mean biologically?

Freshwater ecosystems are a mosaic of sunshine and shade, temperature, and currents. Snails tend to prefer shaded locations, perhaps to avoid predators, exposure to ultraviolet radiation, or reduce temperature fluctuations. Maybe all of the above. Who knows? Early in the life of the cercaria, the best strategy may be to get somewhere else, and anything other than up is irrelevant. Once the cercaria nears the surface and away from the snail from which it escaped, search everywhere: light or dark. As time passes, more time searching shaded areas is the best strategy for finding a suitable host to infect and continue the life cycle. These data support our earlier finding of a higher infection rate of snails in the dark.

Rose did an excellent job. While we were working together, I asked what she was planning to do after graduation. Rose shared she and her boyfriend were heading for California. There was no hint of why or what she would do when she got there. Good to her word, after graduation, Rose disappeared, and I haven't heard a word from her since. I wish her well.

Larval Trematodes — Is Dispersal a Thing?[69]

When cercariae escape from a snail, they typically infect a different host species, sometimes another invertebrate, sometimes a vertebrate, depending on the food preferences of the animal playing host to the adult worm. They have no choice but to swim away from their birthplace, searching for a different host to infect and form a metacercaria. Most fail and die. But others, including *E. caproni*, have a choice. They can either disperse and "hope" to find a different snail to infect or turn around and form a metacercaria in the snail they escaped from. As Dorothy said in the *Wizard of Oz*, "There's no place like home!"

Parasitologists recognize there are plusses and minuses to both strategies. Dispersal is fraught with danger. Cercariae are tiny, aquatic ecosystems large, suitable hosts rare, and predators abound. Becoming lunch is the end of the game. Cercariae don't feed, so energy is a limiting factor. They could run out of fuel in a futile search for a suitable host. The benefit of dispersal is if several cercariae do infect a distant host, they are probably the offspring of unrelated parents. If a rodent living on the shore eats an infected snail, the adult worms swapping sperm aren't clones of each other, which would increase the genetic diversity of their offspring.

If all cercariae exiting a snail take the easy route, infecting the same snail, they risk killing the host or inbreeding with their clones when the snail is eaten. On the other hand, the risk of predation or not finding a host is less problematic. The question I was interested in examining was simple: do *E. caproni* disperse?

None of the previous experiments answered the question. I needed a new setup to determine if cercariae preferentially left the site where they first entered the ecosystem when emerging from the snail. A funnel trap (a variation of the hoop net) might solve the problem. I needed a miniature version cercariae could easily swim through in one direction but not reenter and return to their starting point. I needed a small funnel that would fit inside the PVC tubing used in the previous experiments.

I searched all of the storage and prep areas in Biology with no luck. My next thought was Chemistry. The Chemistry Storeroom was the fiefdom of the irascible, always funny Peggy, or more formally, Mrs. Miller. I went downstairs; Chemistry is on the 1st floor, and Biology is on the 2nd. I told Peggy what I needed. She got up, went into the back, and emerged with a box of nearly perfect-sized funnels. They were just a hair too large to fit inside the PVC tube. I asked if I could have a few. Peggy told me to take as many as I wanted.

I took the funnels and a length of PVC pipe home and got my electric drill and some sandpaper. I inserted the funnel stem into the drill's chuck, tightened it down, and pulled the trigger. I applied the edge of the funnel to a piece of coarse sandpaper and watched the diameter of the top of the funnel narrow. In less than a minute, the funnel fit perfectly into the PVC tube. I polished the edge with fine sandpaper and used a hacksaw to remove the stem. I applied a bead of silicone sealer to the edge of the funnel and inserted it into the tubing. After a few hours, the sealer was dry, and I had a small funnel trap. I made nine more. When inserted into a slip coupler, the device was perfect. The slip coupler formed the lower (or source) chamber holding the cercariae and was separated by the funnel trap from the upper (or dispersal) chamber (Figure 9). Cercariae in the source chamber could swim up following the funnel's sloping walls and pass through the narrow opening into the dispersal chamber. While it would be possible for them to swim back down into the source chamber, the probability was small. I put it on the list.

Francesca Gifford was a gregarious redhead from New York. We met during her first year in an Introductory Biology lab I occasionally taught. She was fun to talk with, and we chatted in the hall during the ensuing years. Fran (her preferred name) registered for Parasitology and claimed the dispersal project for her senior research experience. I explained the basic premise, and Fran got to work searching the literature and writing a review. Based on what she learned and my previous experience, we included light and dark as variables to determine their effect on dispersal. We decided on separate light and dark trials with a snail in the source chamber to give the cercariae an incentive to stay home.

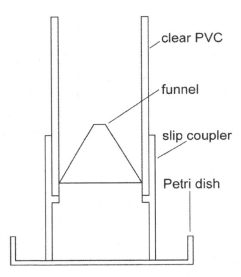

Figure 9.   Dispersal chamber from: Platt, T.R., F. Gifford, and D.A. Zelmer. 2016. The role of light and dark on the dispersal and transmission of *Echinostoma caproni* (Digenea: Echinostomatidae) cercariae. *Comparative Parasitology* 83: 197–201. (Illustration by D. A. Zelmer. Reprinted with permission.)

We collected cercariae and placed them in our dispersal device's source chamber, attached the PVC tube with the funnel trap, added some water, and waited. In lighted conditions, 98% of cercariae left the source chamber. Results in the dark were identical. We repeated the experiment, but we placed an uninfected snail in the source chamber before attaching the PVC tube. The results were similar but with a slight twist. In the dark, nearly all the cercariae migrated to the dispersal chamber, very few stayed in the source chamber, and about 1% infected the snail. With the addition of an overhead light source, close to 80% of the cercariae dispersed, and 20% infected the available snail.

These results were puzzling. Despite their ability to discern light and dark conditions, cercariae are, from a sensory perspective, deaf and blind (and dumb too, as you already guessed). Cercariae do possess the capacity to detect chemicals in the water, but only if they are within a millimeter or so of the source (i.e., a snail). Why would light override their inherent negative geotaxis, resulting in the infection of snails at the bottom of the water column? Perhaps light reduced the swimming speed or increased turning behavior resulting in more opportunities to contact an available snail. I don't know, and now that I am retired, someone else will have to figure it out.

Francesca got a job with a medical therapeutics company following graduation.

17. Interregnum

How you feel about home after a holiday says a lot about home.
— Joyce Rachelle

The turn of the millennium was transformative. I earned the rank of full professor, and a second sabbatical was on the horizon. I was gathering information for the chapter on the TBFs, and working with students on their projects. One other issue loomed large. Our department chair was stepping down, and I was next in line for the job. I never aspired to any administrative post, but this was a rotating position, and to refuse it would merely put the burden on someone else. I wasn't thrilled, but I would likely have to take one for the team.

I chose to apply for my sabbatical for the fall semester, 2000. It would be a "hunker-down-and-stay-in-the-lab" sabbatical, no collecting trips to exotic locales. I still had material from Australia to finish; the book chapter would take time regardless of how well I thought I knew the TBFs, and I wanted to teach Parasitology, which I offered in the spring. I enjoyed the non-majors course, but if given a choice, worms trumped general biology every time.

As soon as I received the letter granting the leave, Kathy (my lovely wife) inquired, "Where are we going?" Kathy knew a reprise of Australia was not in the offing, but she had her own ideas. Our kids were in college, and our second dog (a purebred Keeshond named Cady) died the year before. I don't want to seem insensitive, but both of those things were true.

I loved Cady (Shunda passed away years earlier), but the circumstances brought to mind the following joke:

> A priest, a minister, and a rabbi were debating when life begins. The priest says, "That's easy; life begins at conception." The minister disagreed and shared, "Life begins at quickening when you can feel the baby move." The rabbi thought for a moment, shook his head, and said, "Life begins when the last kid is in college, and the dog dies."

I didn't say it was a good joke, but in Kathy's mind, nothing was stopping us from taking a couple of weeks for ourselves. She asked me where I would like to go and what I would like to see. "Nowhere and nothing" was not going to cut it. I had never been across the Atlantic, and my family roots are in England, so I suggested a trip to the UK. I threw in Down House, Charles Darwin's residence, one of the few places I truly wanted to visit.

Kathy is a born planner. Give her something to organize, and while she will never admit it, she is in hog heaven. We settled on early October for the possibility of weather neither too hot nor too cold, and it would encompass our 27th wedding anniversary. She routinely snuck places she wanted to see into our dinner conversation, and I added the Natural History Museum as a possible destination. The trip began to take shape. The exact itinerary shifted and morphed in its particulars; we would arrive in London for a few days, take the train to Edinburgh, drive through the Scottish Highlands and stay with some folks in Oban. Then we would go to Manchester to find my grandfather Samuel Platt's birthplace in Ashton-under-Lyne. From there, we would drive to the western tip of Wales and catch the ferry to Ireland. After a whirlwind tour of the Emerald Isle, return on the ship, take a train to London and home. We would do this all in two weeks with one carry-on bag each.

During the summer, I concentrated on several papers describing new genera and species of trematodes collected in Australia and the first draft of my chapter on the key to the Spirorchidae. I passed on attending the annual parasitology meeting in Puerto Rico as being too expensive in light of our planned excursion to Great Britain and Ireland. Besides, I didn't want to take time away from the research projects I needed to complete, as promised in my sabbatical proposal. Some of this work stretched into 2001, but I think I more than justified my time away from the classroom.

In early October, we landed at Gatwick Airport and took the train to London. Our accommodations were better suited to graduate students than a full professor, but neither Kathy nor I cared about luxury hotel rooms. After all, how much time do you spend in the hotel when traveling? The room had space for a bed you could walk around if your back was flat against the wall. The bathroom was down the hall and up a half-flight of stairs. Not the Four Seasons by any stretch of the imagination.

Our first stop was the Natural History Museum, the crown jewel of British biology. David Gibson, Head of the Worms Division, met us at the entrance for a behind-the-scenes tour. Despite our disagreement over Spirorchiidae vs. Spirorchidae, David was unfailingly gracious. We met the other members of the parasitology group, had tea, and talked worms for a bit. David took us into the bowels (no pun intended) of the facility and showed us specimens collected by Darwin on his circumnavigation of the globe on the HMS Beagle. After two hours of incredible hospitality, David indicated he had a meeting to attend and turned us loose in the museum proper. Kathy took my picture next to Darwin's statue, and we spent a couple of hours taking in the exhibits.

The next day, we toured Westminster Abbey. Neither of us is religious in the traditional sense; this was another homage to Darwin. Darwin requested burial in the churchyard near his home in Downe, 20 miles southeast of London. Darwin's friends felt his stature in the scientific community deserved more substantial recognition and lobbied for his internment in Westminster with Sir Isaac Newton and other luminaries. They succeeded, and we were in search of Darwin's resting place. The Abbey is genuinely spectacular, regardless of one's religious predilections. We wandered the myriad rooms looking at the crypts of royalty and literary and scientific notables. We finally located Newton's tomb, topped by a statue of the physicist and inventor of my nemesis, "the calculus." Darwin's was a more modest slab on the floor. Calculus is deemed difficult but mastered by high school kids around the world. Natural selection is simplicity itself, yet misunderstood by intelligent people who should know better. Go figure.

Kathy is a whiz at directions and took control of the "underground" map as we toured other locations in London. I love riding subways almost anywhere. Trains offer the best people-watching opportunities in any city. Rich and poor, elegant and avant-garde, are all mashed together with the same goal — to get somewhere else. I find those underground rides more fun and educational than looking at cathedrals, castles, and temples of bygone eras.

Our day trip to Downe and Darwin's home was everything I had hoped. The weather was beautiful, and the train ride delightful. Downe was still the small, rural enclave Darwin inhabited during the middle chunk of the 19th century. A few shops and homes graced the short walk from the train station to Darwin's immaculately restored residence. While grand in comparison to the surrounding dwellings, the house was modest in appearance and appointment. The main structure is a three-story affair with additions Darwin built as his family grew.

I marveled at his study filled with books, bottled specimens, and a microscope. It was difficult to believe his treatise on barnacles, let alone *Origin of Species,* was born in the clutter of such a small space. Kathy and I spent several hours examining the rooms where Charles and Emma tutored their children. The "Sandwalk" was of particular interest. We trod the same path Darwin used as a place to work out thorny issues with his research. I wish I could say I had an 'ah-ha' moment while retracing Darwin's footsteps, but it is more about the person than the place.

The train to Edinburgh was breathtakingly banal. A far cry from my excursion lying on mailbags from Youngstown to Cleveland. We don't (but should) have anything comparable at home — fast and efficient ground transportation. We did the standard tourist stuff; the castle, high tea at a swank hotel (money trumps jeans), and headed northwest to Oban. We almost skipped Scotland until friends Barb and Ian Carmichael convinced us the highlands were a "must-see" on any visit to the UK. Ian, a native Scot, is an internationally recognized chemist who heads the Department of Energy lab at Notre Dame. Ian grew up in Glasgow, and Barb is a champion of anything Caledonian. Barb convinced Kathy if we didn't tour the highlands, we might as well stay home.

Kathy stumbled across something called the Hospitality Exchange, an early iteration of Airbnb. For a $25 fee, you received a list of people willing to host guests for a night or two in exchange for a meal out, and your name was added to the list to reciprocate. She tried London with no luck but fixed on a couple in Oban: Sue and Ned Rimmers. After a few e-mail exchanges, they invited us to spend two nights on Kerrera, an island approximately half a mile off the coast, in a house built circa 1700. We arrived at the ferry landing after a drive through the heath and moors from Edinburgh, parked the rental, grabbed our carry-ons, and hopped on the ferry to the island. There were no cars on our island destination, only bicycles and ATVs. Ned met us at the island landing with an ATV for our bags, and we proceeded up the rutted dirt road to our lodgings for the next two nights.

The house was a mixture of eras; the low ceilings and doors suggested an earlier time when people were either shorter, or energy efficiency was more of an issue. However, the modern kitchen and satellite dish were definite hints these folks had most of the amenities of contemporary life.

We arrived in time for dinner. Sue was busy in the kitchen when the ATV pulled up with our bags and us a few steps behind. After a round of introductions and a quick trip to the 'loo,' we were put to work setting the table for the evening meal. My heart sank when Sue proudly announced the evening's repast was haggis, the Scots' traditional dish. I had heard stories of odds and ends of a sheep stuffed into the animal's stomach and boiled. Mine did a silent turn. We assembled at the dining room table, the organ in question was slit open, and Sue scooped a generous portion of the contents onto my plate. After everyone was served, I took a bite. It was nothing more than sausage sans casing. All those horror stories were wrong; I liked it. I really liked it.

The next day we toured the island on foot. Kerrera is approximately four miles long. We walked from Ned and Sue's home to the coast crossing a verdant countryside dotted with more sheep than people. After lunch, we took the ferry back across the channel and toured Oban and the surrounding country. McCaig's Tower (or Folly) is the most intriguing landmark, a Romanesque building started by one of Oban's leading citizens in the late 1800s as a monument to his family. Mr. McCaig died before completion, and it now stands in mute testimony to his vision or vanity; take your pick. We took our hosts to a local restaurant for dinner as thanks for their hospitality. We returned to Kerrera for conversation and a wee dram of Scotch before retiring for the evening.

The next morning we crossed the channel, retrieved our rental, and headed south for Manchester and Ashton-under-Lyne. I had extensive experience driving on the left side of the road during our time in Australia, so the trip to Manchester was uneventful. We arrived in Aston-under-Lyne at dinner time. After the terrible hotel in London and an early 18th century home in Scotland, I wanted good-old, American-style lodgings. The Best Western on Ashton Road was perfect. We checked in and went looking for a place to eat. We landed at the Dog and Partridge, one of the few places open on Sunday evening. It was packed, and we waited at the bar until a table opened up.

Since we were searching for my grandfather's birthplace, I asked the bartender for the local phone book. He handed me a slim volume covering the northeastern suburbs of Manchester, which included Ashton-under-Lyne. Platt is neither a common nor uncommon surname. In a mid-sized city in the United States, I expect to find between 10 and 15 possible relations listed. That small slice of the UK was home to 119 Platts! I was astounded.

Once seated, I was hoping for some information to answer a riddle I had been unable to solve. My grandfather's birth certificate contained, in addition to Ashton-under-Lyne as his birthplace, the designation "Botany." It might be a postal zone, but I had no idea. An official-looking gentleman, whom I presumed was the owner of the Dog and Partridge, approached our table, and I posed the question. He rubbed his chin for a moment and shared his wife's family had connections to Botany, and they were dining at the table right behind us. Following a round of introductions, we were given directions and told when we saw the Napolean Inn, we were in the heart of Botany. As far as our hosts were concerned, Botany was a neighborhood. If it meant more in the mid-1800s, they had nothing to offer.

The next day we followed the directions and located the Napolean. The surrounding area appeared to be the type of working-class neighborhood my grandfather might have called home over a century earlier; streets lined with brick rowhouses as far as the eye could see. The only difference I could imagine was the exterior brick was probably much cleaner than at the height of the Industrial Revolution. We didn't try to contact any of the multitude of Platts in the phonebook. While the evolutionary history of worms is of great interest to me, my genealogy is not. We got back in the car for the drive across Wales to Holyhead and the ferry ride to Ireland. Wales was beautiful in an empty, green, and rock-strewn kind of way. In the late afternoon, we arrived at Holyhead and checked into a lovely B&B with a spectacular view of the harbor.

The three-hour ocean voyage from Holyhead to Dún Laoghaire was smooth, and we headed for the rental car kiosk to get our transportation for the next few days. The agent presented us with a vehicle that appeared to have survived a demolition derby, but just barely. It was also the only vehicle available — take it or leave it. We took it and headed north and west to Dublin and on to Kilkenny. We were lucky enough (wink, wink) to hit Dublin at rush hour. Stop and go traffic and waiting for several light cycles to cross a busy intersection was the norm. At one light, I was startled by a knock on the driver's side window. I rolled it down only to be informed by the motorist behind me the brake lights on our rent-a-wreck didn't work. I made the rest of the hour and a half drive through the dark and stormy night (it was dark and raining), petrified someone would slam into the rear end of the car when I stopped at an intersection.

We arrived at the B&B in Kilkenny, bone-tired and hungry. We asked our hosts for directions to a pub and headed back out into the night. The establishment (whose name vanished in the mists of time and an aging brain) was lovely; a large bar, dark wood, and a few regulars hoisting their pints. Exactly how I pictured an Irish pub. The fare, limited to various meat pies and stew, sounded wonderful. I am no connoisseur of beer, so my strategy was to order the most popular brew at any particular establishment. I asked. The waiter answered, and his reply almost knocked me off my chair. The best-selling beer at this classic Irish pub? Budweiser! Budweiser is one of two beers I detest (the other is Iron City if you must know). I requested the name of the runner-up and received something local and drinkable.

The next day we took in ruins, castles, and other sites that bore me to death and headed back to Dún Laoghaire and Dublin. Kathy loves bus tours. They are an excellent way to see and learn a lot about a place when time is limited, and we only had the afternoon. After seeing the sights of British-Irish conflict written on walls pockmarked by bullets, we headed for the Temple Bar area for more traditional tourism: shopping and eating. I love the latter and am uninterested in the former.

The restaurant we chose was a perfect tourist haunt. The downstairs was full of folks bending their elbows while we climbed the stairs to the second-level dining room. The food was excellent and the entertainment better. A local musical group played traditional Irish folk songs accompanied by two talented (by my standards) dancers. We had an authentic River Dance five feet from our table. The next day we boarded the ferry for Holyhead, caught a train to London, Gatwick, and home.

We remained members of Hospitality Exchange for several years with no inquiries. South Bend was not a must-see destination for world travelers.

I spent the remainder of the semester in my office and lab working on various research projects. It was also clear our chair would resign after one three-year term, and somebody would need to step up. Almost everybody expected that somebody to be me. I didn't want the job, but I wasn't concerned about taking it either. We had an incredibly congenial group. Everyone got along, and we never had any difficulty reaching consensus on departmental matters.

I assumed the mantle of Department Chair in June. The outgoing Chair, Dick Jensen, my companion teaching non-majors biology, was taking a sabbatical leave in the fall. Dick assumed the responsibility for finding a replacement to teach his Ecology class as many of our majors needed it as a graduation requirement. The young man Dick recommended was a PhD student from Canada and had experience teaching at our branch campus of Indiana University — IUSB. I'll call him Jake. I do remember his name, but it wouldn't be fair to divulge it without his permission. Jake's letters of recommendation were superb, and he presented himself well during his interview. I was delighted not to have to deal with a hiring decision as my first order of business. I was so wrong.

The 'Jake' who arrived at the end of July to prepare for the fall semester was Mr. Hyde to the Dr. Jekyll we hired in May. Summer Jake was aggressive, demanding, and overbearing. I quickly realized we made a horrible mistake; however, we needed to offer Ecology, and it was too late to find a replacement. His contract was only three months, and I thought we could survive almost anything for a semester. His behavior, however, became more erratic. Several of the female staff indicated Jake made them uncomfortable. One woman told me she received sexually explicit e-mails from our new hire. I documented the complaints, got copies of the e-mails, and headed for the Dean's office. The Dean agreed to cut Jake loose. My mind goes from zero to worst-case scenario at warp speed. Due to his erratic behavior, I envisioned Jake going "postal" when he learned of his dismissal. I asked our head of security to escort him off campus and bring his gun just in case.

I learned later Jake was bipolar, which he controlled with medication. He was on his meds when he interviewed in May. At some point, before his July arrival, Jake decided he didn't need them, which resulted in the downward spiral in his behavior. Jake never returned to campus, nor do I know how his life unfolded after our unfortunate encounter.

Although I enjoyed being Chair, it did require more time than I initially anticipated; research took a backseat to teaching and service. Beginning in 1977, I published at least one paper a year except for the two years (1986/87) following my termination from the University of Richmond. The streak ended in 2003. I didn't have any publications in 2004 or 2005. Abandoning research is easy to do. Getting back to it is much more challenging. I got back on track and continued to publish regularly until retirement and for a couple of years afterward. My colleagues recommended me for a second term as Chair. I was granted a third sabbatical after stepping down at the end of 2007. Kathy and I prepared for another worm-hunting expedition.

18. Are You Smarter than a Trematode? II — Adults

Science is not a collection of facts; it is a process of discovery.

— Robert Zubrin

Adult Trematodes — Crowding Effect[50]

Examination of the crowding effect in adult *Echinostoma* started a phenomenal run of student publications involving adult worms in mice. What is the crowding effect? As population size increases, competition for resources increases, and the size of the competing organisms decreases. This is well established in the ecological literature across a range of organisms. The crowding effect was demonstrated in parasitology in the early 1960s using the rat tapeworm, *Hymenolepis diminuta*. I was unaware of any similar studies using trematodes and thought it would be interesting to explore.

I listed two projects on the crowding effect using mice and chicks as the definitive host. *Echinostoma caproni* develops into an adult in both. Students who chose the projects could pick one host or the other and infect one group of animals with a small number of metacercariae and the second group with a larger number. They would care for the animals for approximately a month, kill them (I still don't use the term sacrifice), remove the worms, and following standard procedures, stain them and make slides for later examination. The students would measure a number of structures from a sample of worms collected and statistically compare the measurements from each group for differences. Each student would need to make at least 20 measurements on a large number of worms (30–50 from each group) for a meaningful statistical analysis. Making slides and measuring would be tedious and time-consuming. I needed students who were dedicated and meticulous. Lindsay Stillson and Erin McQueen, best friends and outstanding students, claimed the projects.

This study would entail an extensive time commitment involving animal care, slide making, and measuring specimens. I felt the summer would be the best time to complete these projects. Saint Mary's sponsors summer research awards, SISTAR grants for student/faculty collaborations. I felt we had a good chance of receiving one of the four awards made each year. The grant provided $3,500 for the faculty member and an equal sum for the student. Faculty are not required to work during the summer and, hence, not paid. Therefore, the stipend is a salary supplement for working during a period not required by contract. For the student, it replaced money they might earn through a summer job. We prepared and submitted the application. I received a call from the Grants Committee Chair informing me the grant could only cover one faculty member and one student. There was no provision for funding two students.

Since I worked every summer without being paid anyway, I suggested using my stipend to support the second student. The answer was a flat, "No!" The committee couldn't set such a precedent. I was dumbfounded. What kind of precedent? Putting the education of our students first? The chair was adamant: one student and one faculty member. It was *Sophie's Choice*: Lindsay or Erin. I explained the problem, and they dropped the application. They were best friends, and neither was willing to sell out the other. In hindsight, I should have submitted the proposal with one student. If we were successful, I could have paid the second student myself. I am embarrassed to admit I didn't consider that end-around at the time.

Lindsay lived in town, and Erin was an hour away. We worked out a schedule where I would do the day-to-day animal care, I would be in the lab anyhow, and they would come in to do the necropsies, slide-making, and worm measurements. Before the end of the semester, we infected two groups of mice and two groups of chicks with 10 and 25 metacercariae, respectively. There was one wrinkle. Our mouse supplier almost always gave us an extra animal, and I made an executive decision. We infected the last animal with 300 metacercariae. They went home and waited for the infections to mature.

The summer proceeded as planned. Lindsay and Erin returned to campus, dissected their animals, and made slides before classes started. They carved time out of their schedules to do the measurements. My approach to student work is for the student to do something and check their work. Repeat the process several times until I was satisfied they were doing the procedure correctly, and leave them alone. Both made 1,000 measurements: 20 per worm on 50 specimens. While Erin did an excellent job and earned departmental honors for her effort, technical issues precluded publication of her work.

Statistical analysis of Lindsay's worms showed differences in some of the structures. The overall picture suggested individuals from the 25 metacercariae infections were smaller than those from the ten metacercarial infections, and worms from the 300 metacercarial infection were smaller still. While suggestive, the analysis was not as convincing as I had hoped. We needed something more definitive. I suggested Principal Component Analysis that Dick Jensen and I used to solve the *Aptorchis aequalis/A. anfracticirrus* problem (Chapter 13). At this point, I would have asked Derek Zelmer to perform the analysis and discuss the results with us. However, Lindsay was taking Dick's Biostatistics course that included PCA. Lindsay completed the analysis herself, and the results were startling. Worms from each infection group clustered together, meaning they were more similar to each other than they were to the worms from the other groups.

There was a significant crowding effect on *E. caproni*. The differences were so distinct a person not familiar with the worms might conclude each of the three clusters were separate species, not all *Echinostoma caproni* from a single source! Lindsay presented the results of her work at the Annual Midwest Conference of Parasitologists in June and won the Raymond Cable Award for best undergraduate student presentation. Her work was accepted for publication in the *Journal of Parasitology* the following year, with Lindsay as the first author. Lindsay completed the DVM program at the Purdue School of Veterinary Medicine and is a practicing veterinarian. Erin is a physician. They were two of the brightest and most talented students I had in my 28 years at Saint Mary's.

Adult Trematodes — Because I was Annoyed[53]

Lindsay and Erin's work stimulated me to think about more projects involving adult worms. One issue, in particular, gnawed a bit. Lindsay infected mice with 10, 25, and 300 metacercariae. The mice infected with 10 and 25 metacercariae yielded approximately 70% of them as adults, but only 7% of the 300 metacercariae infection produced adults at necropsy. We hypothesized an immune response to the massive infection resulted in the low numbers, but we didn't know. Bernie Fried published a paper suggesting some possible reasons for our low yield. Most of them suggested we did something wrong. I was determined to test the hypothesis of immune rejection, and we knew what we were doing. I infected 21 mice with 300 metacercariae each. I necropsied three animals at 1, 4, 8, 12, 16, 20, and 24 days post-infection to track what happened over time and see if the number of *E. caproni* declined and when. All of the mice yielded 150–250 worms at each interval, and there was no sign of massive worm loss. So much for the immune rejection idea.

I experienced a problem determining the position of the worms in the intestine during the infection. Bernie established a protocol dividing the small intestine into five equal sections for locating the worms to infer location and/or movement over time. The more I looked at my data, the more I thought Bernie's approach was too coarse-grained and didn't reflect the small intestine's functional divisions. I read anatomy and physiology texts and medical tomes on gastroenterology. The small intestine is divided as follows: 0–7% duodenum; 7–45% jejunum; and the rest, ileum.

The duodenum receives the acid chyme from the stomach, which mixes with sodium bicarbonate from the pancreas to neutralize it before further processing. Bile is added from the gallbladder to emulsify fat, and multiple enzymes flow from the pancreas to initiate the chemical breakdown of proteins, lipids, and carbohydrates. Absorption of nutrients begins there as well. Most of the enzymatic processing of food and nutrient absorption occurs in the next section, the jejunum. The last half of the small intestine, the ileum, resorbs bile salts and returns them to the liver for conversion back into bile, absorbs some water, and whatever nutrients manage to get past the jejunum. The ileum is a food desert as far as the worms are concerned. Bernie's first 20% contained the duodenum and part of the jejunum, the second 20% was all jejunum, the third 20% part of the jejunum and anterior ileum, and the last two sections were all ileum. I thought a more fine-grained division might yield better information.

Simply measuring where each worm was in the gut wouldn't work because small intestine length varies from mouse to mouse. I devised the following scheme. I removed the small intestine during the necropsy, placed it in cold saline, and stored it in a refrigerator. The treatment shrank the gut as much as possible and standardized the measurement of total length. Before examining the intestine for *E. caproni*, I placed it on a dissecting pad and pinned both ends while avoiding stretching, and measured the length to the nearest millimeter. As I opened the intestine longitudinally, I pinned the gut to the pad to prevent stretching. When I found a worm or cluster of worms, I placed a smaller pin to mark its location and removed and counted the worms. Once I finished, I measured the position of each worm or worm cluster to the nearest millimeter using the small pins as markers. Finally, I divided the gut into 20 equal segments based on total length and placed the worms in the section (or sections) indicated by the pins. If a cluster of worms spanned two segments, I split the total between them. This procedure provided a standard method for locating *E. caproni* in association with the small intestine's functional structure.

Adult Worms — Migration[57]

Now let's get back to Anna Kammrath. When I realized the light/dark behavior study was an impossibility (Chapter 16), she needed something else to do. I didn't have any ideas for other projects involving cercariae, so I devised an experiment using adult worms. In the late 1960s, several parasitologists reported the rat tapeworm, *Hymenolepis diminuta*, undertook a diurnal migration in the rat intestine. Put simply, the worms moved anteriorly and posteriorly in the gut on a 24-hour cycle. As far as I was aware, there was nothing in the literature describing a similar phenomenon for trematodes.

As I envisioned the experiment, we needed to dissect mice at eight points over 24 hours to detect a migration if it happened; more than I could reasonably expect of one student. Anna was on board, and I enlisted the next student who came to talk to me about supervising her senior research. I pitched the idea to Emily Graf. She listened as I provided the background information and the experimental design. Emily thought for a minute and signed on.

One issue had to be addressed. The project required dissecting mice at four times during the day and four times during the evening and early morning. Staying up all night wasn't an option, so we put the evening mice on a reversed light/dark cycle. Instead of having the lights go on at 7 am and off at 7 pm, they came on at 7 pm and off 12 hours later. We could do all of our work during the day.

I ordered the mice, and we situated them in the animal room: four groups of three animals on a standard light cycle in one room and the other four groups on a reversed light cycle in a separate room. We gave them a week to acclimate and infected each mouse with 20 metacercariae of *E. caproni.*

The infection process is relatively simple. You need a large (one gallon) wide-mouth jar with a layer of cotton on the bottom, ethyl ether, a Pasteur pipet with a thin plastic tube attached, and some metacercariae. We dissected snails previously infected with *E. caproni* and removed the metacercarial cysts. We placed 20 cysts in a small amount of saline in the wells of a depression plate and carefully sucked them into the pipet. We dropped a mouse into the jar containing a small amount of ether. Mice react to ether by scratching their nose with their front paws and are unconscious in about 30 seconds. The mouse was quickly removed from the jar and held by the scruff of the neck while the tube was inserted into its mouth and down the esophagus. Squeeze the rubber bulb on the pipet, and the metacercariae were in the stomach. The mice recovered in less than a minute, no worse for wear. In 20 years of doing experiments on *E. caproni*, not a single mouse died during the infection process. The method is fast, effective, and humane.

The experiment provided the first evidence trematodes undergo a diurnal migration. The worms moved to the anterior portion of the ileum during the late evening and early morning. In the late morning and early afternoon, they retreated to the posterior part of the gut. As any home-owner who has experienced a mouse infestation knows, mice are active at night and rarely seen during the day. They eat while we sleep and process food during the day while sleeping, and we are busy with our lives.

Echinostoma caproni tracks the feeding behavior of their host. This worm is a mucosal grazer. It feeds on cells lining the intestine that are continually sloughed off and any food reaching the ileum. Like other para-sitic flatworms, trematodes possess a unique outer covering, or tegument, capable of absorbing small molecules from the environment and using them as an energy source. The worms compete with the intestine, as well as their fellow travelers, for nutrients. Moving anteriorly in the gut as food approaches puts the worm in a better position to compete for whatever is coming its way. The worms move posteriorly as the food enters the ileum allowing them to bathe in the digesta as long as possible. Very cool indeed!

Anna and Emily disappeared into their adult lives, and I have no idea of the direction they took.

Adult Trematodes — The Next Step[65]

What stimulates *E. caproni* to move? Food seemed to be the obvious answer, but what is obvious isn't always correct. It seemed reasonable to suggest *E. caproni* moved in response to food, but will they move in response to nothing, saline, or a simple nutrient like glucose passing through the intestine? Those questions were the foundation of the most ambitious experiment I conceived for a senior research project. Testing two variables (time and food type) would require 36 animals and more time than I could reasonably expect from one student. I listed it as a dual endeavor. Arianna Rodriguez and Guadalupe (Lupita) Quintana, best friends, were game.

The experimental design required 36 mice divided into four groups of three: no food, saline, glucose, and standard rodent chow. Animals from each treatment group would need to be examined 1, 2, and 4 hours after being given their post-fasting meal; four treatments × three intervals × three animals at each interval, for a total of 36 animals. All mice were infected with 20 metacercariae and allowed access to food and water *ad libitum* (i.e., anytime they want it) for 28 days. Prior to necropsy, all animals were fasted for 24 hours and then given access to one of the four post-fasting regimes. Ari and Lupita dissected the mice and froze the intestines, as described in Chapter 14.

The absence of food or saline does not stimulate the release of gastric or pancreatic secretions or bile. There should be no advance warning to the worms nutrients are on their way because there isn't any food in the gut. The worms didn't move. They were in the same place at 4 hours as they were at hour 1. The worms in mice fed glucose moved a little anteriorly, but the worms in mice given food containing the full range of nutrients jumped forward by a bunch. *Echinostoma caproni*, like *Hymenolepis diminuta*, seemed to detect the presence of food coming their way and responded by moving in the direction of the oncoming meal.

Migration was an intriguing find, but the real question was, "What are the worms (cestode or trematode) responding to?" They seemed to move before the food actually arrived, so it must be a chemical signal enticing the worms to respond. For nearly three decades, scientists tried a range of food components (carbohydrates, lipids, and proteins) with no response from *H. diminuta*. Finally, Mike Sukhdeo, of Rutgers University, cracked the problem.

The answer turned out to be simplicity itself — peristalsis. When a complex of nutrients pass from the stomach to the small intestine, it stimulates the release of hormones, and other chemicals, initiating contraction of the smooth muscle surrounding the gut. These contractions, or peristalsis, move the food posteriorly for digestion and absorption. Peristalsis is not stimulated by the absence of food, the presence of saline, and only mildly by glucose — hence, we saw no movement in worms in any of our controls. Peristalsis is the most reliable indicator food is on its way, and parasites (*H. diminuta* or *E. caproni*) move forward to get their share of the bounty. While we did not directly test for the influence of peristalsis on *E. caproni*, if it worked for cestodes, it seemed like the best bet for *E. caproni* as well. Arianna completed a PhD in microbiology at Northwestern, and Lupita is pursuing a PhD in Public Health.

19. The Art Exhibit*

The first mistake of art is to assume that it's serious.

— Lester Bangs

In late April 2007, I received an e-mail from Krista Hoefle, a member of our Art Department and Director of the Moreau Galleries at Saint Mary's. Krista inquired if anyone in our department had drawings or photographs suitable for display in the Moreau's opening exhibition for the 2007–08 academic year. I couldn't think of anyone offhand, but I dutifully forwarded her note to the other members of the department. Nobody responded to my inquiry.

I produced illustrations, India ink on vellum, of the worms I described for scientific publications. Drawings, or photographs, are required to show the features presented in the body of the text. The quality of illustrations appearing in science journals ranges from professional — some folks have grant money to hire artists to do this work or are themselves talented — to those more typical of a grade school art project. I fall somewhere in the middle. I never took an art class or worked with anyone knowledgeable in the craft to teach me how to draw my subjects. I learned by trying to reproduce what I judged to be good drawings from the literature at my disposal. I would never compare my illustrations to the best, but they are far from the worst.

---

*Portions of this chapter appeared in TAXONOMY AS ART — The delicate dance of scientific realism and abstractions of reality. *The Journal of Parasitology Newsletter. Vol. 30. No. 3. 2008* — © American Society of Parasitologists 2008. (Used with permission.)

I debated whether to share my efforts with Krista. My drawings were functional — what I saw through the lens of a microscope. They are not art. Throwing caution (and my sense of dignity) to the wind, I replied and attached a file containing a scanned image of *Buckarootrema goodmani* (Digenea: Pronocephalide) from a 2001 publication in the *Journal of Parasitology*. I told her it was representative of the drawings I produced, and if she was interested, I had more. Before I left for the day, Krista replied — Yes! We made arrangements for her to come to my office to look at the originals.

A few days later, Krista was in the Science Hall, and I had a dozen or so "worm" drawings spread out on the floor. Krista is an elfin presence with a perpetual smile and more energy than any adult should rightfully possess. She was excited and wanted to use my drawings as part of the August opening. Moreau Gallery is a forum to display the work of students, faculty, and invited artists. Krista left and promised she would be in touch during the summer to select the "pieces" for display. I was sure she would think better of her rash decision, so I kept mum about the prospect of this event taking place.

In mid-June, Krista was back and chose a dozen of my drawings for the opening. I also provided her with copies of the papers in which each illustration appeared. Because the originals varied in size and condition (some were 15 to 20 years old), Krista decided to reproduce them using a high-resolution scanner and print them on art stock of uniform size. Krista raised the question of a name for the exhibit. She suggested something like *Weird Bugs*. I demurred. First, they weren't bugs. Bugs are insects in the Order Hemiptera. Second, if this was going to happen, I wanted the opportunity to educate anyone interested enough to take the time to look at them. I suggested the title *Taxonomy*. I wrote a brief description providing an introduction to the discipline, the drawings, and how I produced them.

Taxonomy

Taxonomy — "The theory and practice of describing, naming and classifying organisms." Lincoln, R.J., G.A. Boxshall, and P.F. Clark. 1982. *A Dictionary of Ecology, Evolution and Systematics.* Cambridge University Press, Cambridge, p. 298.

Taxonomy is the foundation of biology. Every organism must have a unique binomen consisting of a genus and species name rendered in Latin. This practice dates to the work of Carl von Linné, better known as Linnaeus, in 1758. Current estimates of the number of extant (currently living) species range from 10–100 million. Of these, approximately 2 million have been formally described. The goal of the current biodiversity movement is to describe the remainder before they are overtaken by extinction.

A formal publication includes a written description and illustrations that provide characters differentiating the new species from all others described. The figures may be photographs or line drawings, but they must show the relevant features that enable future workers to identify the species at a later date. Line drawings are generally considered superior to photographs because they tend to demonstrate the critical features more clearly.

The illustrations in this exhibit are of parasitic worms (Phylum Platyhelminthes) collected from the circulatory and digestive systems of turtles in different parts of the world. The size of these animals ranges from approximately 1 to 3 mm (0.04 to 0.1 inches) in length.

The specimens were obtained by careful dissection of a turtle and collecting the worms found. The worms must be killed and fixed to prevent decay — embalming of sorts. The worms were stained to show their internal structure and mounted on microscope slides for examination.

Drawings were done initially in pencil with the aid of a *camera lucida* attached to the microscope. Extreme care was taken to accurately depict the general structures of the organism and those specific features necessary to differentiate it from all other forms of life. The pencil drawing was rendered in India ink on vellum. In short, I did a lot of tracing! The written description and illustrations were submitted to a journal for consideration for publication. The journal's editor sent the manuscript to several experts in the field for comment. If the reviews were favorable, the paper was accepted for publication. Once published, the name becomes part of the scientific literature and recognized as a validly named species.

Thomas R. Platt, PhD, has been a member of the Department of Biology of Saint Mary's College since 1986. Tom obtained his BA from Hiram College (1971), MS from Bowling Green State University (1973), and PhD from the University of Alberta (1978). He is the author, or co-author, of approximately 50 peer-reviewed papers and is considered an expert on the trematode family Spirorchidae. He has no training in art, formal or informal as if you couldn't tell, and from time to time writes poetry — or some semblance thereof.

Tom would like to thank Krista Hoefle for thinking that what he does is worth looking at and giving him the opportunity to share his work with people who would never willingly peruse an issue of the *Journal of Parasitology*, *Comparative Parasitology*, or *Systematic Parasitology*, the journals in which most of this work was published.

Over the years, I dabbled in poetry, as do many, many people. Most are for my wife for birthdays and anniversaries. Some are good and some banal, but they are always appreciated. Some deal with science, and I had four poems published in *Perspectives in Biology and Medicine* (University of Chicago Press) when they still published what they termed "that challenging medium." I am no more a poet than I am an artist. I don't work at poetry. I don't set aside a portion of each day to write and revise. Typically, something pops into my head I can't let go of, and it eventually finds its way onto a piece of paper and into a digital file. I thought I might be able to express the process of my work more clearly in verse than prose, so I wrote four poems providing a chronology of worm hunting, taxonomy, and biological nomenclature. The following poems accompanied the exhibit.

The Game

It may hide in the blood or the gut or the lungs.
It does not want to be found.
Discovery is death.
Natural selection prepared it
To hide from antibodies and
  Killer T-cells,
  Enzymes and
  Macrophages.
But scalpel and forceps strip away
  layers of skin and muscle,
To lay bare my quarry,
For fixation in formalin,

Staining and mounting in balsam,
Microscopic examination.

But the rules of the game are clear.
I must know what I found,
I must give it a name.
I must share its identity with all who
    want to know.
It is the least I can do.

What's in a Name?

Taxonomists,
    foot soldiers of biology.
Revealing
    the diversity of life on Earth.
Describing
    things no one has seen,
Giving names,
    identity.
And the names drive you crazy.
Forgotten Latin from Mrs. Wilson's 8th-grade class.
Why would anyone call something
    — anything — *Griphobilharzia amoena*?
But names can be as revealing as
    the life they describe,
A beautiful mystery — if you wish.

It Is Not Art

It is not art,
    what I do.
I draw only what I see.
I draw without interpretation
    or creativity.
I draw only what they need to see.

A shadow on the cave wall,
    in crisp black and white.

The essence of the ideal transferred
    to paper by my unsteady hand.
The ideal, its essence extracted, mounted
    on a slide,
    tucked safely in
    a museum drawer.

Someday, someone will open the drawer
And attempt to reunite essence
    and ideal.
If I have done my job,
She will see what she needs to see.
What I saw, and
    the two will fit perfectly.
The shadow will prove Plato's thesis.

But it is not art.

Science and Stamp Collecting

Doyens of the laboratory,
Clad in perfect white lab coats
Makers of hypotheses and experiments,
Call it stamp collecting —
    not Science.
And yet,
When they report their results of
    experiments on
    Schistosoma mansoni, or
    Mus musculus, or anything else
They must attach a name,
    a stamp,
To validate their Science.

It is fair to say my work does not rise to the level of Maya Angelou, Billy Collins, or Mary Oliver. I do not expect a call to serve as Poet Laureate of the United States.

Once Krista had what she needed, I put the whole business out of my mind. I had to prepare for the upcoming semester. I was still Department Chair and had to deal with last-minute administrative issues and plan a sabbatical in Malaysia at the beginning of the New Year. I didn't have time to think about the event until it was right on top of me.

Exhibit openings are typically scheduled for Friday evening from 5 to 7 pm. I had two lectures plus other chores requiring my presence until late afternoon. I climbed the steps to the gallery a few minutes before 5. My family and a few friends joined the students and faculty who regularly attended these events. Wine, cheese, and crackers occupied the center of the room and garnered most of the early attention. I guess folks need some fortification before undertaking an evening of viewing art, or they wanted to get theirs before the goodies disappeared.

My drawings lined the walls between the entryway and the refreshments (Figure 10). I stopped for a moment to marvel at what was happening. People who weren't parasitologists were taking time out of their day to examine my drawings. I greeted the folks I knew and got a large glass of red and a hunk of cheese. I had two hours to go.

I shared the opening with Dominic Paul Moore (Chicago) and Paul Campbell (New York City). They had an exhibit entitled "Profile Me" based on the FaceBook and MySpace phenomena and paintings by Gianna Commito, an artist from Kent State University. I spent two surreal hours drinking wine (which reduced my anxiety), explaining to the small cadre of students, faculty, and visitors what they were looking at, the use of a *camera lucida* to produce the drawings, and why on earth anyone would do this kind of thing. I had fun explaining parasitology and why it was important. I hope I raised the consciousness of at least a few of the attendees. I also enjoyed talking to "the real" artists. I tried to convince them what I did was not art, while they tried to convince me it was. The debate was a draw, and I have to rate the evening as one of the more surreal in my then not-quite 60 years.

One of the most gratifying outcomes of this experience wasn't people telling me they "loved" what I did (although who doesn't like a bit of stroking, sincere or not). It was the fact that colleagues with whom I worked for

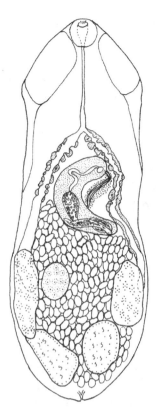

Figure 10. ***Buckarootrema goodmani*** from: Platt, T.R. and D.R. Brooks. 2001. Description of *Buckarootrema goodmani* n.g., n.sp. (Digenea: Pronocephalidae), a parasite of the freshwater turtle *Emydura macquarii* (Gray, 1831) (Pleurodira: Chelidae) from Queensland, Australia and a phylogenetic analysis of the genera of the Pronocephalidae Looss, 1902. *Journal of Parasitology* 87: 1115–1119. (Reprinted with permission.)

over two decades discovered what I did when I wasn't teaching or doing requisite committee service. Most faculty have little idea what their counterparts in other departments do in the way of research. My "art show" served as a catalyst for conversations that would have never happened otherwise.

I wrote a short piece for our society newsletter to encourage members of the American Society of Parasitologists to do the same. I asked, "Could you have an art show? Why not!" Many of the ASP members who do taxonomic work or captivating photography are far more talented than I. Most liberal arts colleges have art departments and galleries with limited budgets. Not all gallery directors might be as adventurous as Krista, but I think many would welcome the opportunity to display the images produced by their colleagues in the sciences.

The show closed on September 26th. My career as an artist was over. Krista gave me the prints she made for the exhibit, and the show lived on, for a time, in the hall outside my office. I had fun talking with students who stopped to check out my drawings.

I had one inquiry from a colleague in the math department about obtaining a print. I was happy to oblige. She received one of the prints in exchange for a dollar, and I could lay claim to being a professional artist. The bill is in a frame on my wall. I never reported it to the IRS.

20. Heroism in Student Research — Trematode Eggs[64]

A hero is somebody who voluntarily walks into the unknown.
— Tom Hanks

I had notions for research projects dealing with diurnal (daily, or about 24-hour) rhythms and *E. caproni*: escape of cercariae from the snail and release of eggs by the adult worms. However, I was too old to stay up for 24 hours, and I didn't like asking students to do things I wouldn't do myself. I included both of them on the list with little expectation any student would express interest in either project. I was wrong again. Gabrielle (Bre) Hussey appeared in my office after Thanksgiving and wanted to discuss the egg production project. I gave her a broad-brush description of what would be required, collecting mouse poop every two hours for 24 hours and processing it to estimate the number of eggs present. This wouldn't be a one-time thing. She would have to repeat the process on the same mice at least three times. Bre was nonplussed. Infecting mice and collecting poop was simple; isolating and counting eggs were another matter.

Collecting poop, while not difficult, does pose some technical problems. Mice are typically housed in cages lined with wood chips, which is not conducive to collecting the hundreds of fecal pellets these animals produce. My solution was to cut the bottom off of four standard cages. I made a trip to the hardware store and purchased a few yards of wire mesh with ¼ inch openings. I cut the mesh a bit larger than the open bottom of the cage, cut and folded the corners, and glued it to the open bottom. The poop could pass freely through the mesh. Now we needed something to catch the stuff. I had an idea what we needed and where to find it — Gordon Food Service (GFS). GFS sells bulk food and a range of items for serving large groups. The lids for their large aluminum baking trays would do the trick. The lids were large enough to cover the area of the cage bottom, and when placed on the shelf immediately below the cage caught all the poop passed by one mouse. Unfortunately, all the pee and food crumbs landed there as well. Bre would have to clean the fecal pellets of any food to obtain an accurate weight for her calculations.

A typical mouse produces about 1.3 to 1.8 grams of poop a day when food is continuously available. Something I am sure you wanted to know. Bre would be collecting between 0.1 and 0.2 grams every two hours. She needed to collect, clean, and weigh the droppings of four mice. Once removed from the tray, the pellets were weighed, transferred to a centrifuge tube and softened in a dilute solution of sodium hydroxide, crushed and mixed, forming a suspension. The suspension was passed through a layer of cheesecloth to remove large debris and centrifuged at low speed for three minutes to concentrate the eggs. Bre removed the bulk of the supernatant (fluid in the tube), leaving 2 mL of fluid containing the eggs. The liquid was well mixed, and three 100-μL (or 0.1 mL) samples were removed with a micropipetter. Bre transferred the fluid to a microscope slide and added a 50 mm × 22 mm coverglass. She counted the eggs using a compound microscope at low power (40×).

Every two hours, Bre would collect, clean, and weigh the feces from four mice. Process the feces, make 12 slides (three for each of the four mice) and count the eggs on each slide. We did a couple of practice runs to estimate the time required. After three trials, Bre did everything in about 1 hour and 45 minutes. Enough time to run to the bathroom, if necessary, and start again. She was still game.

We infected the mice with 25 metacercariae and settled on 28, 51, and 58 days post-infection for the marathon poop collections. The first was standard for the development of mature infections of *E. caproni*. The other two were more matters of convenience based on Bre's work schedule and family obligations during the summer. I assisted during regular work hours to give Bre a bit of respite and make sure everything was running smoothly. Overnight, Bre was on her own.

At the end of Bre's three 24-hour marathons, we killed the mice to document the number of worms present and calculate the number of eggs produced per worm and the number per mouse each day. For three of the mice, Bre found a nearly linear increase in egg output: 10,000 to 15,000 on day 28; 28,000 to 60,000 on day 51; and 30,000 to 40,000 on day 58. Egg output in two mice rose linearly over the three collection periods. The third mouse mirrored the same pattern for days 28 and 51 but declined on day 58. The fourth animal was a puzzle. Egg output was similar to the other three animals on day 28 but dropped to less than 1,000 on days 51 and 58, respectively. The first three mice harbored 22–26 *E. caproni* each, while the laggard had only two. Were there only two the whole time? Were worms lost after 28 days? We don't know. *Echinostoma caproni* tend to cluster in the intestine, enhancing opportunities for sex. The two worms in the fourth mouse were not found near each other and may have relied on self-fertilization, depressing egg production. We don't know that either.

Despite the small number of worms in mouse four, the pattern of egg production and release were sufficiently similar between the mice each day, and between days, we combined the data for analysis. Fecal output followed the pattern of mouse activity: more poop at night and early morning when the mice are active, and less during the day when they are sleeping. The animals released the highest number of eggs from 8 pm to 8 am and significantly fewer during the daylight hours. We suggested the daily pattern of fecal production and egg release enhanced the dispersal of parasite eggs in the environment. When the mice were out and about, exploring their environment for food and mates, they pooped more. They spread *E. caproni* eggs to locations where they had a higher probability of entering aquatic ecosystems necessary for the miracidium to develop and eventually infect a snail. Many eggs would be lost, but the release of thousands of eggs every day should ensure some reach places they could develop and continue the life cycle.

Bre initially considered graduate school in research, and I wrote enthusiastic letters of recommendation. Before graduation, she had a change of heart and planned to apply to programs in physical therapy. I retooled my letters and sent them off. I can't imagine she wasn't accepted, but I never heard one way or the other. It happens.

21. The 30-Year Project[61]

> *It's not that I'm so smart, it's just that I stay with problems longer.*
> — Albert Einstein

The hair on the back of my neck goes up whenever I enter someone's office and see a poster with a quotation by Albert Einstein, implying a shared genius with the occupant. I think it is the height of arrogance to compare oneself with a *bona fide* genius unless you are a physicist. If you have a degree in physics, you are exempt. I am comfortable using the epigraph heading this chapter because it is about persistence, not intelligence.

Obtaining a PhD is not a test of intelligence but persistence. Don't get me wrong; PhDs are, on the whole, bright — some more than others. However, there are lots of smart people without advanced degrees. Being smart isn't enough. You have to be able to put up with long hours, low pay, and an uncertain future at the end. You have to want it, and you have to have the persistence to finish. A fair number of people don't. The biggest stumbling block is the thesis. Most of us know someone with an ABT — all but thesis. They do the coursework but can't muster the wherewithal to complete the research or sit down and write it up. They lack persistence. I frequently describe myself as not that smart but persistent. This is a story of persistence.

Westhampton Lake separates Richmond College from Westhampton College. The two colleges initially established for men and women, respectively, constitute the University of Richmond (Chapter 8). When I arrived in 1978, the lake was chock-a-block with turtles and still was when I was summarily let go. I decided turtles would form the focus of my research. It didn't take long to start catching the dominant species present: snapping turtles (*Chelydra serpentina*) and stinkpots (*Sternotherus odoratus*). Both housed a cornucopia of worms.

Part of my strategy for necropsying a turtle involved removing the digestive tract, esophagus to anus, in one piece. Removal of the intestine included the skin surrounding the anus to ensure I didn't miss something at the terminal end of the rectum. I separated the esophagus, stomach, small intestine, and large intestine and put them in individual dishes to avoid mixing parasites from one organ with those from adjacent sites.

My interests were with worms in the gut and other internal organs. Lots of things colonize the turtle shell (think barnacles on whales) which, while interesting to some folks, are not to me. You have to draw the line somewhere. One afternoon, during the summer of 1981, I was dissecting a stinkpot turtle and opened the rectum. I cut through the anal sphincter, a band of circular muscle that keeps the poop from indiscriminately falling out of the body. The final cut included the small circle of skin surrounding the anus or the perianal (around the anus) region. I've already told you more than you want to know, but there is more. The skin around the anus of stinkpot turtles is a bit different from most other turtles I dissected in my career. In most turtles, the skin is rough, typical for turtles, but flat. In stinkpots, the skin forms a series of peaks and valleys that expand as the feces pass out of the rectum. Why the difference? I have no idea. These peaks and valleys are called perianal folds and ideal places for something to hide.

As I examined the rectum's terminal end, I stretched the perianal folds and noticed something wiggling. There were tiny nematodes (round-worms) inhabiting that unlikely location. I grabbed a pipet and sucked some up, and deposited them on a microscope slide for a closer look. I wasn't sure what they were, but I was intrigued. I collected and preserved as many as possible and continued to collect them from any infected stink-pot having the ill-fortune to wander into my nets.

A few weeks later, I decided to take a stab at identifying these denizens surrounding the bunghole. While the worms were small, they required treatment with a clearing agent. Clearing renders the organism translucent and the internal organs visible microscopically. Nematodes demonstrate a tube-within-a-tube design. The worms are long and slender, and their internal organs are mostly long and thin. At first blush, nematodes are boring, boring, boring. However, numerous small differences permit separation into various levels of classification. In the early 1980s, the Phylum Nematoda consisted of two classes, the first major taxonomic division, the Phasmidea, and the Aphasmidea. The Phasmidea contained most of the parasites of animals and few free-living and commensal species. The reverse was true of the Aphasmidea: few animal parasites and lots of free-living and commensal forms. I quickly realized these worms belonged to the Aphasmidea, which was a problem. Most parasitologists, myself included, are familiar with the Phasmidea, but the Aphasmidea were as foreign as space aliens. I decided to press on. I worked them down to the superfamily Monhysteroidea and the family Monhysteridae. I knew I was in over my head and needed help. I needed someone who specialized in the other side of the nematode divide.

I met Virginia Ferris at a parasitology meeting a few years earlier, and we struck up a friendship. Virginia was on the faculty at Purdue University and was an expert in free-living nematodes. I figured she would be able to tell me what these things were in short order. In the pre-internet area, I wrote her a letter and waited for a reply. Virginia responded quickly and volunteered to take a look. I boxed up the perianal worms and shipped them off to west-central Indiana. Within a few weeks, Virginia shared news, both disappointing and exciting. My worms didn't look like anything with which she was familiar (disappointing); however, they might be new (exciting). Virginia shared she planned to attend an international nematode conference (they do exist) in London and asked if she could take them with her. Perhaps someone there might shed some light on their identity. What could I say but yes? I had never been to London, but my worms were on their way.

The news from London wasn't encouraging. Upon her return, Virginia told me nobody she spoke to had heard of nematodes living in such circumstances. She also shared she made an executive decision and gave the specimens to Michael Baker of the Muséum national d'Histoire naturelle in Paris. I knew of Michael, but we had never met. He was the wunderkind of parasitic nematodes. Mike obtained his PhD from the University of

Guelph with Roy Anderson, my sometime nemesis regarding my PhD work on *Parelaphostrongylus*. So they hopped the channel — something I'd never done. I never heard from Mike, but I learned my worms had been repatriated a year or so later and were residing at the Smithsonian. These well-traveled specimens made two transatlantic crossings while I didn't have one to my name.

Mike sent the specimens to Duane Hope, senior nematologist at the Smithsonian, about 100 miles north of Richmond. My worms traveled 8,000 miles only to land 100 miles up the road. I learned Duane had them from Duane. His initial examination suggested a new species and a new genus, and possibly a new subfamily. Certainly good news. He would do the work, and I would get my name on the paper. Fine by me.

Initially, I wrote to Duane every few months to find out what was happening. The intervals increased. Over the next six years, I was turned down for tenure at Richmond, hired by Midas Rex, and moved on to Saint Mary's. Other projects, more important than mine, required Duane's attention. He was appointed Chair of the Department of Invertebrates and had to cut back on research, and so on. Finally, more than two decades after my specimens landed in our nation's capital, Duane informed me he was retiring. What did I want him to do? Exasperated, I told him to send everything back. I would figure it out.

Sometime later, I struck up a conversation with Scott Gardner at the American Society of Parasitologists' annual meeting. Scott is the director of the Harold W. Manter Museum of Parasitology at the University of Nebraska, Lincoln (UNL). Scott loves nematodes, and our conversation segued to the stinkpot worms. Scott was interested and thought he could find a student willing to take on my oddly located nematodes for a Master's project. Scott would supervise, and I thought, why not? Scott said he had a student in mind and thought she would do a good job. Great! This thing might happen after all. I sent the worms to UNL and put them out of my mind.

A couple of years went by with no word. Scott was an infrequent attendee at the ASP meetings as he spent his summers in South America collecting parasites from native rodents. Finally, he informed me the student left for medical school without doing anything, let alone a "good job" on her Master's. I was back to square one. Once again, I fell into a funk, wondering if this reluctant debutante would ever be introduced to the scientific community.

More than a few folks work on free-living and commensal nematodes; I didn't happen to know any, and Virginia Ferris had long since retired. I had an epiphany. I did know someone who worked on this kind of nematode. At least I thought I did. Maybe. During my time in Alberta, there was a young Indian woman, an undergraduate student, who worked in the lab — Jyotsna Sharma. I was pretty sure she had done a PhD in Belgium on free-living nematodes. I hadn't seen or spoken to Jo in nearly 30 years, but it was worth a shot.

Sometimes the internet is nothing short of magical. A quick search on Google Scholar revealed a J. Sharma active in taxonomic studies of free-living nematodes. The search took about five minutes. In another five, I had the office phone number of Jyotsna Sharma at the University of Texas, San Antonio. I picked up the phone, dialed, and she answered. Our 30-year divide melted away like the Wicked Witch of the West. We spent 15 or 20 minutes catching up; school, jobs, family, and life in general. I got down to business. I gave Jo a thumbnail history of the worm, its extensive travels and asked if she would be interested in completing the journey. It was an immediate yes! I felt this was the end or at least the beginning of the end. I was always more comfortable working with people I knew; you understand their underlying character. I felt good when I wrote to Scott and asked him to send the slides and vials back to me, relabeled the box, and mailed it to San Antonio.

Jo enlisted the help of Eyualem Abebe of Elizabeth City State University in North Carolina and Manuel Mundo-Ocampo of the Unidad Sinaloa in Mexico, both outstanding nematologists and friends of Jo's. From then on, things moved quickly. Jo was my go-between and kept me up to date regarding progress on various fronts. In less than a year, the manuscript describing *Testudinema gilchristi** appeared in my inbox. I read it carefully, noting a few grammatical and other minor issues

---

* *Note: Testudinema* (*Testud* = turtle, *nema* = nematode) *gilchristi* (Dr. Willie J. Gilchrist, the ninth Chief Executive Officer, Elizabeth City State University, for his continuous support of nematological research at the institution.) Dr. Abebe proposed the name, and I assume Dr. Gilchrist was pleased with the honor.

requiring attention. The paper was submitted to the journal *Nematology* for review and was accepted. The final version appeared in print near the end of 2012, a little over 30 years from the first time I opened the perianal folds of a stinkpot turtle and observed the wiggling worms who called that unlikely location home.

22. Malaysia

The traveler sees what he sees. The tourist sees what he has come to see.
— G.K. Chesterton

My second term as Department Chair was coming to an end, and I was eligible for a third sabbatical for the 2007–08 academic year. I wanted to make one last collecting trip in the hope of finding new TBFs (turtle blood flukes — it's been a while). The question was, where? India was a possibility. Numerous genera and species occurred on the subcontinent, but Indian parasitologists were unwilling to loan material for study. Collecting at least some of them for more careful morphological examination and molecular analysis would be extremely valuable; however, I had no contacts in India.

Sam Loker, University of New Mexico, did extensive work in Kenya on schistosomiasis, the human blood flukes responsible for more than 200 million infections worldwide. Sam invited me to tag along on one of his visits, but the more I read about the unrest in the country and the level of lawlessness, the less enchanted I became. The possibility of being carjacked by folks armed with automatic weapons did not appeal to my risk-averse nature.

One of the people Mark Rigby recruited to describe the two new Australian nematode species was a possibility: Reuben Sharma of the Universiti Putra Malaysia (UPM). I knew nothing about him beyond his contribution to our paper and an article he wrote redescribing a related nematode species from turtles in Malaysia. I contacted Mark and asked for his thoughts on approaching Reuben about a possible sabbatical visit. Mark gave me a thumbs up and suggested I contact him. I did.

Malaysia was a mystery to me beyond the Petronas Towers in downtown Kuala Lumpur, featured in the film *Entrapment*, starring Sean Connery and Catherine Zeta-Jones. I laid out my aspirations for the trip. I wanted to collect and examine turtles for parasites, specifically TBFs. I anticipated a two-month stay beginning in January 2008. Reuben's response was enthusiastic. He would love to have me visit, provide laboratory space, and assist in collecting turtles.

The question was how to make this adventure a reality. I estimated the cost of the excursion to be around $5,000, too large for Saint Mary's but too small for most granting agencies. The trip would probably be my last hurrah in field collecting. Kathy and I had always been frugal, and we could afford the expenditure. We paid for it ourselves.

The internet had blossomed in the 15 years since our Australian adventure, and I found a website run by expatriates based in Kuala Lumpur. It wasn't for me. I would be in the lab, but Kathy needed things to do while I was working, and she had tons of questions. Her first concern was how to dress and behave in a predominantly Muslim country.

Minna Schwarz-Siem, the wife of a Ford Motor Company executive and former librarian, was delighted to learn of our visit and was a font of information about life in southeast Asia. E-mails flew back and forth between Kuala Lumpur and South Bend. At one point, Mina shared that she and her husband, Daryl, spent time in the Peace Corps after college. To make a long story short, they served in Fiji with friends of ours from Hiram, Chuck and Betty Pritchard, who were posted there simultaneously. There's the small world thing again!

The flight was a horror, over 20 hours, 16 plus to Seoul, and another six to Kuala Lumpur. We arrived in the middle of the night and couldn't expect anyone to meet us at two in the morning. After deplaning, clearing customs, and immigration, we found a nearby hotel, but our heads didn't hit the pillow until around 5 am. Reuben picked us up four hours later. We were shell-shocked, sleep-deprived, and whisked off to Tesco, a British version of Costco, to pick up supplies. We had no idea what we might encounter in our new digs, and therefore no idea what we needed. We filled a shopping cart with guesses and headed for our new home.

Kathy and I hoped to rent a house or apartment within biking distance of the campus to simplify transportation. Reuben rented a three-bedroom condominium for us in a gated community approximately six miles from UPM. As we passed the guardhouse at the entrance of Country Heights, we were taken aback by the row of McMansions (including one belonging to a former prime minister) lining the left side of the road as we wound our way to our condo. I started to wonder about the cost but was too tired to care. We made plans for dinner with Reuben and Sumita, Reuben's fiancé, and as soon as they departed, we both fell asleep.

Apparently, we sublet from some expatriates who were on holiday in Europe. The condo was beautiful, with three bedrooms, two bathrooms, a spacious living area, and a balcony. We had access to a clubhouse with exercise facilities, a pool, and meticulously landscaped grounds. They left a few utensils, plates, and one small frying pan (or Reuben and Sumita bought it for us), and not much else. The lack of implements would test Kathy's cooking prowess. Fortunately, there was a restaurant in the clubhouse, a five-minute walk from our lodgings we frequented to the point of committing the menu to memory. The food was quite good, and we could both have dinner for around $4 each, and no tipping!

Reuben, and by extension, the university, were incredibly accommodating. They provided a private office complete with a computer and faster internet access than I had at Saint Mary's. I did not learn the condo's cost for nearly a month despite my repeated inquiries about paying for our lodgings. Initially, Reuben informed me the rent was $700/month, which was at the high end of our budget. I was puzzled when he told me I would pay later, but I had no recourse other than to wait. Near the end of the first month, Reuben informed me the rent was actually $950/month, but the university would pay for one month. I assumed Reuben was reluctant to reveal the true cost until UPM coughed up the cash. I wonder what would have happened if they hadn't?

Reuben was a rising star in the School of Veterinary Medicine, and with Sumita, they were a scientific power couple. Both were bright, ambitious, and productive. Reuben's teaching load was heavy. He was being drawn into more and more administrative chores and away from the laboratory.

Reuben had a small staff managing his laboratory for teaching and research — Mai, Rashid, and Ujang. Mai was married, with three children, and wore traditional Muslim garb: long, flowing, floor-length wraps and a headscarf, and always very colorful. Rashid was my best bud. He was outgoing, curious, and never at a loss for words, although I was frequently confused at what he was trying to say. Ujang remained a mystery for my entire stay. He either didn't speak English, his English was limited, or he was not enamored of me. We rarely interacted.

Islam is the dominant religion, which was evident in women's attire. Most, but not all, wore the hijab, although I learned it was a recent development. A decade earlier, the hijab was a rarity, while in 2008, most Muslim women covered their heads, befitting the spread of more conservative religious attitudes in the country. The headscarves were in sharp contrast to the western blouses (or t-shirts), skin-tight jeans, and extreme high heels that made up the rest of the wardrobe of many young women.

Since Reuben was typically busy with teaching and committee work, Rashid was my go-to-guy for anything I needed in the lab. There was little he couldn't scrounge up on short notice. Rashid was either naïve, immature, or a combination of both. He expressed curiosity about all things American and never hesitated to strike up a conversation or ask questions. I loved him for it.

Transportation turned out to be an issue. I couldn't justify purchasing a car for a two-month stay. The distance from Country Heights to UPM was only six miles. Six miles may sound like biking distance, but it wasn't. Two limited-access highways barricaded those six miles. As the old saying goes, "You can't get there from here." At least not on a bicycle. Our new home, surrounded by highways, was as isolated as the island Tom Hanks inhabited in *Castaway*. Kathy and I were prisoners.

For the first month, Reuben picked me up in the morning and dropped me off at the condo on his way home. I think this started to wear on him, although he never gave any outward sign of tiring of my company. During the second month, he arranged for a university driver to do the job. A car appeared each morning and whisked me off to the lab. I dissected a turtle (or two), had lunch, examined another turtle (or two) in the afternoon, and went home. When wrapping up for the day, someone in the lab called the carpool, and a driver arrived fifteen minutes later to take me back to Country Heights. If there was a cost involved for this service, nobody shared it with me.

For more mundane tasks like shopping or sightseeing, we were on our own. Daryl and Mina were occasional tour guides as they had a car and knew the area; however, Daryl was busy closing the local Ford plant before their departure to the States. Our options were local taxis and trains. We arranged cabs to the train station through the clubhouse at Country Heights and then hopped the train to Kuala Lumpur. On the way back, we had to hail a taxi on our own. Malaysia claims the language of business is English. I beg to differ. Most shopkeepers and folks on the street spoke rudimentary English at best.

Our first attempt at procuring a taxi unaided occurred when returning from our inaugural grocery shopping trip. We hailed a cab, loaded our bags, and quickly discovered the driver spoke no English (and our high school Spanish and French were useless). The ride was a lesson in frustration and fear until we providentially spied Country Heights. We gesticulated wildly and repeated "Country Heights" until our driver caught on and delivered us safely home. From then on, we tested the driver's language skills by asking a series of questions designed to ferret out one who spoke passable English before engaging his services.

Catching turtles turned into buying turtles. I anticipated doing fieldwork, going out in the bush, setting traps, and checking them periodically. Reuben informed me he would purchase the animals from licensed trappers. I was surprised and disappointed, but Reuben explained turtle populations were small, scattered, and trapping would not yield the number of animals I hoped to examine in a reasonable time frame. In the absence of fieldwork, there are no anecdotes of encounters with crocs, snakes, or other adventures to surprise, startle, or otherwise enlighten the biological voyeur.

If we weren't going into the field, I thought at least I could accompany Reuben when he went on a buying trip. I hoped a few long road trips would afford an opportunity to chat and become better acquainted. Reuben always found a way to, very politely, exclude me from those excursions. On one occasion, Reuben told me he would head north the next day to buy turtles. I suggested tagging along, and he agreed. The following day I received word, through Rashid, that Reuben was ill and had to cancel. The next day there were new turtles to necropsy — no excuse, no explanation. I was flummoxed but didn't protest. For whatever reason, Reuben didn't want me along, and I wasn't going to press the issue.

Reuben was an enigma. I never knew what he was thinking or what he expected from me. I sensed an air of disappointment, but I didn't know what I had or hadn't done to warrant the sense of frustration I intuited. My perception was he thought I was a world-renowned taxonomist who would help identify a backlog of trematode material he accumulated over the years. What he got was a narrowly focused, aging parasitologist on his last hurrah before retiring. Our relationship was cordial but distant. Kathy and I occasionally joined Reuben and Sumita for dinner at a restaurant but were never invited to their home.

Work in the lab was, for the most part, steady but not terribly rewarding. The most common turtle, the Asian box turtle (*Cuora amboinensis*), was no stranger to parasitological examination. I found trematodes and nematodes (turtles curiously lack tapeworms), but while many were new to me, they weren't new to science. I was able to identify most to genus, if not species, from the literature. The second most common turtle, the Black Marsh turtle (*Siebenrockiella crassicollis*), yielded similar results, a fair number of worms, but most easily recognized.

My days were pretty much the same as long as there were turtles to dissect. I sat on a stool for the bulk of each day, dismantling turtles and examining their innards with a dissecting microscope. When I counseled students interested in graduate school, I tried to convey one hard truth: they had to appreciate the tedium and repetition of research. If there were no turtles, I stayed at the condo while attempting to avoid day trips with Kathy.

I am a lousy tourist. I have no interest in castles, monasteries, churches, or temples. I view most travel as going thousands of miles to buy somebody's junk and those folks reciprocating. Cynical? Possibly; however, it is my opinion. If you love being herded through airport security, crammed into an airline seat designed for a toddler, led through the streets of some foreign city following a tour guide's finger, and buying their junk — have at it. I don't revel in your enthusiasm for the enterprise.

The only problem is my wife does not share my curmudgeonly attitude. Kathy cajoled me into a day trip to Melaka (or Malacca) when the university closed for a religious holiday. Several tour books touted Melaka as a "must-see." Kathy read one description to me on our return, and I wondered how I could have been there and not been in awe of the experience. To be sure, Melaka has historical significance. The town fathers did their best to preserve some remnants, some scraps, of the past to pull in tourist dollars, but they were also mindful their constituents wanted progress in the form of, well, everything else. So when you look at Melaka, there are temples, old cannons, and the like surrounded by McDonald's, Toyota dealerships, and Holiday Inns. They are just as tacky as we are in so many ways. God love 'em. I just don't want to pay $200 and waste a whole day being herded with the other sheep. Everyone has something to show outsiders, but most of it isn't compelling unless you are compiling a checklist of things you have seen as a gauge of your life.

The only saving grace of travel is striking up a conversation with someone you don't know, be it a local or a fellow traveler, and learning something about their life and circumstances. Our Melaka adventure began with an unlikely group who, on the surface, had little or nothing in common: an older Asian couple and their two adult daughters, a couple about our age, and a single gentleman who sounded British. It turned out the Asian family was Korean. The father was a retired professor who obtained a Master's degree from Indiana University in Bloomington (IN) back in the day and spoke excellent English. His daughters both had Master's degrees; one worked in Hong Kong for Yahoo. I didn't catch what the other one did, but both earned degrees in Melbourne. The single gent grew up in Townsville and lived in Melbourne. We had a lovely time relating our experiences living in northern Queensland. The couple was from Perth but spent time in Melbourne and had friends in Chicago. The world is a small, small place. Everyone had some connection with everyone else, except, as far as I could tell, the professor's wife. If she spoke any English, I never heard it. She never engaged in any of the cross conversations.

On the other hand, I have no problem with Kathy taking trips with girlfriends leaving me to my own devices. She made arrangements for two girlfriends to join her for a nine-day tour of Thailand, Cambodia, Bali, and Singapore. Carol Jackson and Kathy were realtors for the same company,

Cressy and Everett, in South Bend. Barb Carmichael and her husband, Ian, are longtime friends who recommended we tour Scotland on my previous sabbatical. Carol and Barb arrived, took over the condo for a day and a half, and then they left, almost in the same breath. I was utterly and blissfully alone.

Shortly after the women left, I made a discovery in the laboratory that jacked up my heart rate. While examining the lungs of a Black Marsh turtle, I noted something out of the ordinary, something funny (remember Isaac Asimov) — small clusters of debris. I collected a sample for examination with a compound microscope, and what did I see? Trematode eggs! How did I know? First of all, it is my job to know these things. Second, each egg contained a miracidium, the larval stage of those worms. The miracidia were small (about a third of the volume of the egg) and incredibly active. They swam in tight circles in their tiny enclosures, like a hamster in a plastic ball. When transferred to freshwater, the egg burst open in a matter of seconds, and the miracidium took off like a bat out of hell! Where there are trematode eggs, there are adult trematodes.

I cautiously teased the lung tissue apart with watchmaker's forceps and examined the blood vessels with extreme care. Then I saw it, something waving and contracting, definitely not a blood vessel or connective tissue, but a worm. I instantly realized I had something new. All I had to do was collect it — intact. Removing long, thin worms from long, narrow blood vessels is no easy chore; remember *Aphanospirorchis kirki*. I managed to extract an entire worm measuring about a quarter of an inch long and the width of a human hair. Careful examination of the lung of both marsh turtles and box turtles yielded a dozen intact worms (plus fragments) in the next few days. I got what I came for, a new TBF. Little did I know it would be eight years until I introduced *Baracktrema obamai* to the world as a new genus and species.

The ladies returned from their whirlwind tour of southeast Asia thrilled, enlightened, and considerably poorer. During their stop in Cambodia, their tour guide directed them to a jewelry store where they were flattered, cajoled, and scammed. Each of them paid an exorbitant sum for what turned out to be fake gems. Being conned is one thing; having the person you paid as a guide complicit in the fleecing added insult to injury. Kathy spent $400 on a diamond ring for our youngest son when he found the woman of his dreams. It turned out to be worth a buck and a half — maybe. Barb and Carol demonstrated similar acumen with their purchases. Live and learn.

Once her friends departed for the States, Kathy began organizing pictures and writing up her adventure as a homemade blog for folks back home. At one point, I peeked at the computer screen, showing a picture of an old, ornate building. I asked what it was. Kathy replied it was a temple (duh!), but she couldn't remember its name or location. That brief exchange validated my view of travel. If she couldn't remember something she saw less than a week earlier, what was the point?

One of the highlights of their trip was a visit to Angor Wat, an ancient, abandoned temple in the Cambodian jungle. Tourists flock to it annually only to be fleeced by the vendors (of cash) and monkeys who steal anything they can. Shortly after her return, the History Channel featured a documentary on the self-same temple. In an hour, less commercial interruptions, we saw more of the temple and learned more of its history than Kathy did while physically present. I rest my case.

Reuben informed me before we left the States, I would be expected to present a seminar as repayment for their hospitality. I was less than thrilled. I lectured for a living and gave over a hundred ten-minute presentations at professional conferences. The 50-minute seminar is a whole different beast. My work was episodic; a new genus or species, the redescription of a poorly described trematode, or an experiment attempting to answer a very specific question that doesn't lend itself to in-depth exploration. I stored several short talks on my computer in the off-chance lightning would strike once I was there. I was secretly hoping Reuben would forget the whole thing, and I would escape without performing.

The first month passed without mention of a seminar. I began to fantasize my wish might come true. Reuben shattered my dream. My time at UPM was drawing to a close, and seminar slots were limited, so he scheduled it for the day before our departure for a short trip to Penang to visit an old friend from graduate school before returning home.

I had to decide what I could talk about for nearly an hour that might be remotely interesting to the vet students and faculty. After reviewing the contents of my computer, there was only one option. I would repurpose my PhD seminar on *Parelaphostrongylus*. I already repurposed it for my parasitology class and saved it on my laptop. All I had to do was find a hook to link it to Malaysia, and I was home. I gave Reuben the title of "Lungworms, Deer, the Canadian Rockies, and…Malaysia?" Soon thereafter, posters appeared announcing my talk on March 3rd during the hour before lunch.

Species of *Parelaphostrongylus* infect deer and Malaysia is home to two species of cervids; *Muntiacus muntjak,* the barking deer, and *Dama dama,* the fallow deer. The barking deer is native, while the fallow deer is an introduced species. A third, closely related deer relative, *Tragulus javanicus* (the lesser mouse deer), is also native to Malaysia. Due to the pathogenic nature of meningeal worm in anything other than its natural host, white-tail deer, wildlife biologists are incredibly cautious about introducing them into new habitats for fear of decimating the local cervid fauna. No species of *Parelaphostrongylus* or other near relatives were known from the southern hemisphere, so the barking and mouse deer might harbor undescribed species or even a new genus. All I had to do was present the original material and add a few slides illustrating the danger of introducing exotic animals into the region and the potential for new discoveries.

When I entered the auditorium, I was stunned to find I had a full house. More than 150 students and faculty turned up for this well-worn talk, although they didn't know it had been 30 years since I first presented it as part of my doctoral requirements. The presentation was a blur, as they almost always are from my perspective. When I finished, I looked at the clock, and 50 minutes of my life had disappeared in a heartbeat. It took me a few seconds to appreciate the thunderous applause coming from the audience. I may have disappointed Reuben in other ways, but his speaker brought down the house.

Hands went up, and questions started flying. Most were simple and easy to answer. Then disaster struck. A faculty member raised his hand, was acknowledged, and asked a question I did not understand. I am not saying the question was complicated and beyond my comprehension; I had no idea what he said. The words coming out of his mouth made no sense. Many of the faculty obtained their graduate training in the UK and spoke with curious but distinct British accents and used British pronunciations — sort of. I asked if he would please repeat the question. He did, and I was no wiser than before. I apologized and asked again. The embarrassment mounted. The problem was one word I heard as "pree-*day*-toor" with a hard accent on the second syllable. It finally dawned on me; he was saying, "predator!" I made a self-deprecating remark, which initiated a wave of laughter and defused the embarrassment I caused. I apologized profusely and provided an acceptable answer. The aftermath was a bit chaotic, and I received more congratulatory comments than I deserved.

The entire Parasitology lab assembled at the local mall for a farewell lunch. I received some small gifts from Rashid and Mai. I remember them fondly, and I hope they feel similarly. A few days earlier, I was showing Rashid pictures of our home in South Bend. Five years earlier, we moved to a house on a small lake on the city's west side. Rashid studied the scene of our dock, pontoon boat, and the expanse of water. After a few moments, he turned and said, "I will come and fish from your dock." I hope he does. I don't fish, but I will buy a rod and reel and get him a license if he makes good on his promise.

We left for a four-day visit to Penang, the island state of Malaysia, home of Ray Leong, a fellow graduate student from Alberta. Recall it was Ray, who unceremoniously backed into a hole in the ice and the frigid water of Cold Lake in eastern Alberta on my first foray into the field as a PhD student. Ray and I hadn't had any contact since he finished and returned home in the mid-1970s. I tracked him down (again, bless the internet), told him Kathy and I would be in Malaysia, and he invited us to visit.

Kathy booked a room at Lone Pine Hotel. The Lone Pine was lovely in the slightly seedy manner of aristocracy fallen on hard times. Our room was on the top floor overlooking the pool and the ocean, an idyllic location for our last few days in Southeast Asia.

Ray and his wife, Helen, were consummate hosts and tour guides. They treated us to everything I hate, and Kathy loves — temples, museums, and historical landmarks of all types. Despite my aversion to this point-and-look tourism, I pride myself on being a good guest. I smiled and nodded at appropriate moments and occasionally gushed about something that seemed to warrant gushing — and promptly forgot 99% of it. A fun fact: we saw the third longest reclining Buddha in the world. I have no idea where his two longer exemplars might be. If I had a bucket list, they would be on it.

We flew from Penang back to Kuala Lumpur, met Reuben and Sumita at the international terminal, and retrieved the rest of our luggage. They presented us with a large, blue peacock from the Tenmoku Pottery. I am not sure why they thought we were blue peacock people, but we were, at that moment, the proud owners of an 18-inch long ceramic replica and tasked with transporting it halfway around the world without breaking. The peacock survived the trip, our two grandsons (so far, so good), and is prominently displayed in our home.

Earlier in our stay, Reuben and Sumita indicated their fandom for Harry Connick, Jr., who was performing in Kuala Lumpur later in the year. They wanted to go, but the tickets were a tad out of their price range. Kathy and I purchased two tickets to say "Thank you" for their kindness and hospitality. They were overwhelmed. Perhaps their gift was meant to balance the ledger. The peacock is India's national bird, symbolizing grace, pride, and beauty (http://greetingindia.tripod.com/symbols.html). Maybe I was less of a disappointment than I thought. We exchanged farewells and boarded the plane for the 20-hour return to Chicago and 2-hour drive to South Bend.

Malaysia was both a success and a disappointment. Let's start with disappointment. It is clear by this point I don't like to travel. It doesn't mean I don't like being in new places. I enjoy extended stays (e.g., graduate school in Edmonton, our year-long stay in Australia) when I have the opportunity to settle into a routine, go to the grocery store, get to know my neighbors, and learn about the zeitgeist of the place — looking at stuff, not so much.

Kathy and I hoped for a house or apartment in a middle-class neighborhood populated with academics and the like. A place close enough to UPM where a bicycle would suffice for my daily transportation. A place with small shops and restaurants we could frequent and get to know the owners and regular patrons. Perhaps our expectations were unrealistic, and those types of neighborhoods don't exist in the vicinity of UPM or at all. Possibly Reuben thought we needed something more luxurious, and Country Heights was more befitting my status (in his mind) as a Professor at an American college. I will never know.

Country Heights was a beautiful prison, but a prison nonetheless. We were isolated, geographically, culturally, and personally. I needed a chauffeur to get to work. The homes and condominiums were western in style and size, with a few architectural nods to their owners' Indian and Chinese heritage, but few people were out and about. Wealthy people seem to fear those who aren't, and ensconce themselves behind guarded walls and exhibit little interest in their neighbors. It doesn't matter where you are in the world; it seems to hold true. I enjoyed talking with the graduate students and technicians in Reuben's lab, especially the effervescent Rashid. That was as close as I came to learning about Malaysia and Malaysians. I wish I had had the opportunity to expand the circle a bit more.

From a scientific perspective, the trip went as well as I could have hoped. My primary objective was to find new TBFs. I discovered what I would eventually christen *Baracktrema obamai* as a new genus and species.[68] I also found two new species of *Hapalorhynchus*: *H. tkachi* and *H. snyderi*, named for two friends and colleagues, Vasyl Tkach and Scott Snyder, who honored me with the patronym *Choanocotyle platti*.[62]

Then there was the gallbladder worm. Both the box turtle and the Black Marsh turtle had small trematodes in the gall bladder. They also harbored a species of *Telorchis*, a genus of trematodes common to turtles worldwide, in the small intestine. The small gallbladder worms looked like juveniles of the larger *Telorchis* under the low magnification of the dissecting microscope. I wasn't aware telorchiids made a sojourn in the gallbladder before reaching adulthood, but stranger things occur. I didn't think any more about them until our return to South Bend. When I examined the stained specimens under the higher magnification of the compound microscope — they were not *Telorchis*!

Despite my best efforts, I had difficulty placing them. Nothing previously found in turtles looked like these tiny denizens parasitizing the gall bladder. I expanded my search and arrived at the genus *Opisthioglyphe*. Species of this genus infect frogs and turtles but live in the intestine, not the gallbladder. I did what I frequently do in these situations; I contacted someone more knowledgeable — Vasyl Tkach. Vasyl, originally from Ukraine, holds a faculty position at the University of North Dakota. Vasyl is a taxonomist *par excellence*, and more importantly, he wrote the key (identification guide) to the family of trematodes containing *Opisthioglyphe*. I called him and told him what I thought I had. He was skeptical but incredibly curious. I sent him my slides, and he concurred. It was indeed *Opisthioglype* and a new species to boot! He offered to write the paper.[63]

Vasyl had one other question. Due to its unusual location, he wondered how the new species were related to those calling the small intestine home. In addition to being an outstanding morphologist, Vasyl is also an incredibly talented molecular taxonomist. His question was — had I preserved any specimens in ethanol for DNA analysis? I searched all the vials I brought back, and the answer was "No." I stained and mounted all of the little buggers on microscope slides in Canada balsam. Once the worms had been through such an ordeal, the DNA was no longer fit for extraction, sequencing, and comparison with their intestinal kin. Vasyl and other parasitologists have searched the gall bladders of box turtles in various locations in southeast Asia to no avail. The genealogical relations of *Opisthioglyphe sharmai* (named in honor of my Malaysian benefactor, Dr. Reuben Sharma) will remain a mystery for the foreseeable future.

It is difficult to reconcile the death of 56 turtles for a return of maybe 150 vials of worms. Those vials represent half a dozen publications, my curiosity, and ambition. I can convince myself of the value of what I do. I increased our understanding of the diversity of life on Earth. The data may help assess the evolutionary and ecological forces that shaped and are still shaping our planet. But what if I had stayed home? What if I hadn't traveled halfway around the world? Would it have made any difference? Probably not. The turtles would still be dead, cooked, eaten, and perhaps the energy used to make human babies. Would the world be able to stumble forward without whatever information I contributed to the "scientific" literature over the next few years built on the backs of those 56 examples of "God's noblest creatures?" I suspect the planet would be fine or as fine as our abuse of it allows. As I have written elsewhere, the world does not need us and will probably be better off when we are gone.

Once back in South Bend, I began to contemplate the future. By the time I would be eligible for my next sabbatical, I would be 65 — retirement age. Could I imagine doing anything like this again? Could I imagine plunking down somewhere completely foreign for some extended period? Would the bad bits have faded far enough for me to entertain the idea of another collecting trip? The answer was, "No." I killed hundreds, if not thousands of animals (if you include mice and snails) in my career and made the best of every one of them by publishing what I discovered. Killing takes a toll. It is soul-numbing. I didn't know it at the time, but I had dissected my last turtle.

My father imbued in us a respect for people regardless of their station in life. The fourth-floor coffee shop in the Zoology building at the University of Alberta taught me to respect people no matter where they are from. The people at UPM were warm and welcoming, but Kathy and I have been treated similarly almost anywhere we have been. I keep in touch with the Malaysian cohort through Christmas letters to Reuben. He and Sumita married and had a daughter, Elene. Several years later, Reuben asked me to write a letter supporting the promotion of Dr. Rehanna Sani, another parasitologist at UPM, to the rank of Professor before she retired. The promotion would boost her pension substantially. I doubt what I wrote swayed the committee in any meaningful way, but it was an honor to contribute. And she got it!

23. Student Research — Finally, A Taxonomic Study[58]

Accuracy of observation is the equivalent of accuracy of thinking.
— Wallace Stevens

Andrea Firth hailed from my hometown of Youngstown. My cousin was her family's attorney. Small world — again! Andrea started her senior research on a larval trematode/snail project that went bust. Some do through no fault of the student. She invested a great deal of time and energy before realizing no result, positive or negative, was forthcoming. In these instances, students can opt for a paper and presentation of what-ifs: present the hypothesis, the experimental design, a brief overview of why it didn't work (if you can figure it out), and what if it had. It is not the most satisfying outcome, but sometimes it can't be helped. Research doesn't always cooperate with hopes, dreams, or time constraints. I gave Andrea the what-if option, and she declined. She wanted results. Fortunately, Andrea's failure occurred in the spring of her junior year, so I had time to think of something for her to do over the summer.

I didn't have any ideas for interesting *E. caproni* experiments, but I did have a taxonomic study investigating something that annoyed me for years. The Malaysian box turtles harbor a small trematode, *Neopolystoma liewi*, in the conjunctival sacs (i.e., under the eyelids). This worm was described as a new species in 2000 in a paper with Louis Du Preez of South Africa as the senior author. Dr. Du Preez is an excellent and prolific scientist. His primary area of study is parasites of amphibians. As such, our spheres of interest rarely overlap.

Occasionally, Dr. Du Preez ventures into turtle parasites. While I have a great deal of respect for Louis, he does something that infuriates me. When preparing specimens for staining and mounting on microscope slides, he flattens them with a coverglass. Pressure from a coverglass cannot be controlled. The force applied may be a little or a lot. Pressure on a small, soft-bodied animal will distort the animal in unpredictable ways. I learned this procedure in graduate school. The joke at Bowling Green was if you wanted to describe a new species, press down harder on the coverglass! I abandoned the practice decades earlier, and as a reviewer, I criticized anyone using the method, including Louis Du Preez. It never seemed to matter. The same people kept squashing their worms year after year after year.

Most of us kill and fix our specimens in hot fluid, saline, or formalin. Either will produce excellent specimens, with less distortion, for a wide range of trematodes, including monogenes. In all fairness, I understand why Dr. Du Preez opted to flatten his worms. Monogenes have a large posterior holdfast called an opisthaptor. The holdfast of *Neopolystoma* consists of a large disc with six suckers arrayed around the periphery. There is a small hook at the bottom of each sucker and additional small hooks on the disc proper. Flattening the specimen allows better visualization and more accurate measurement of the hooks. The solution to this problem is simple. Heat-kill the worms you are going to measure and flatten a few to observe and measure the hooks. I used the process successfully and recommended others follow suit.

I had enough specimens of *N. liewi*, properly fixed for a redescription of the species. I offered Andrea the project as a replacement for her failed larval trematode work. Andrea, a bit of a stoic, agreed with more enthusiasm than I anticipated.

The project was straightforward. The specimens were already stained and on slides (Figure 11). All Andrea had to do was measure them and compare hers to those in the original publication. She found coverglass pressure increased the measurements of structures used for taxonomic purposes by 30–70% compared to their heat-killed counterparts. Andrea

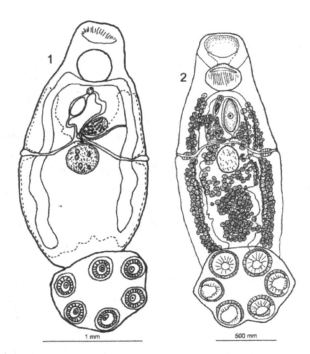

Figure 11. *Neopolystoma liewi* from: Platt, T.R., A. Firth, and R.S.K. Sharma. 2011. Redescription of *Neopolystoma liewi* Du Preez and Lim, 2000 (Monogenea: Polystomatidae) from *Cuora amboinensis* (Testudines: Geoemydidae) with Notes on Specimen Preparation. *Comparative Parasitology* 78: 286–290. 1. Flattened. 2. Unflattened. (Reprinted with permission.)

did an excellent job and produced a clear, concise paper and presentation. The work was accepted and published.

During the spring of 2019, I was perusing the program for the upcoming American Society of Parasitologists meeting scheduled that summer for Rochester, Minnesota. Dr. Du Preez was scheduled to attend, and a face-to-face encounter was inevitable. From the moment I realized that Louis and I were destined to meet, I was nervous about how this encounter might play out. Considering I had taken him to task for flattening his specimens both in reviews and print, an angry confrontation was not out of the question. While waiting for the elevator at the end of a morning listening to presentations of new research, I felt a hand on my shoulder.

I turned, and a friend said, "Tom, there is someone I would like you to meet. Tom, Louis Du Preez. Louis, Tom Platt."

Louis was my height, but 20 pounds heavier, dressed in a black shirt and matching trousers. He was compact and built like a linebacker, perhaps a slight paunch, but not fat. His hands were large with thick, meaty fingers, hands that could crack walnuts. Louis smiled broadly and exclaimed, "I have been waiting to meet you for years." He couldn't have been nicer. We got on famously and had several enjoyable conversations. At the end of our last conversation, just before the meeting ended, Louis took me aside and shared that he no longer used coverglass pressure on specimens used to measure morphological features. Sometimes you can make a difference!

CODA

Of all the things I accomplished professionally, student research is a significant source of pride. I supervised 76 student projects during my 28 years at Saint Mary's. The number, per year, was not exceptional. Everyone in the department carried a similar load. The fact that 15 students contributed data resulting in 12 publications in national and international journals was exceptional. The dedication of the students (or at least the overwhelming majority) was extraordinary. Few of our students pursued careers in research; however, post-graduation surveys reveal the Senior Comprehensive Research experience as the most important and valuable of their time in the department. I think I can speak for all of my colleagues in saying we are proud of them, their commitment, and their accomplishments. It would have been nice to have had more money, more time, and more recognition for our efforts, but watching those young women take the stage and present their research with pride and confidence was reward enough. I don't regret a minute of it.

Publications by TR Platt

(Numbers correspond to text references; * indicates a student publication)

1. Rabalais, F.C., M.L. Eberhard, D.C. Ashley and T.R. Platt. 1974. Survey for equine onchocerciasis in the midwestern United States. *American Journal of Veterinary Research* 35: 125–126.
2. Platt, T.R. 1977. Helminth parasites of two turtle species in northwestern Ohio. *Ohio Journal of Science* 77: 97–98.
3. Platt, T.R. 1978. A report of *Polymorphus paradoxus* (Acanthocephala) in *Microtus pennsylvanicus* from Hastings Lake, Alberta (Canada). *Proceedings of the Helminthological Society of Washington* 45: 255.
4. Platt, T.R. and W.M. Samuel. 1978. *Parelaphostrongylus odocoilei* (Nematoda: Metastrongyloidea): Experimental studies on the life cycle in the mule deer (*Odocoileus h. hemionus*) and other cervids. *Experimental Parasitology* 46: 330–338.
5. Platt, T.R. and W.M. Samuel. 1978. A redescrption and neotype designation for *Parelaphostrongylus odocoilei* (Nematoda: Metastrongyloidea). *Journal of Parasitology* 64: 226–232.
6. Platt, T.R. and A.O. Bush. 1979. *Spinicauda regiensis* n. sp. (Nematoda: Heterakoidea), a parasite of the Ball Python (*Python regius*). *Journal of Helminthology* 53: 257–260.
7. Platt, T.R. 1980. Observations on the terrestrial molluscs in the vicinity of Jasper, Alberta (Canada). *The Nautilus* 94: 18–21.
8. Platt, T.R. and D.B. Pence. 1981. *Molineus samueli* n. sp. (Nematoda: Trichostrongyloidea: Molineidae), a parasite of the badger, *Taxidea taxus*. *Proceedings of the Helminthological Society of Washington* 48: 148–153.
9. *Falls, R.K. and T.R. Platt. 1982. Survey for heartworm, *Dirofilaria immitis*, and *Dipetalonema reconditum* (Nematoda: Filarioidea) in dogs from Virginia and North Carolina. *American Journal of Veterinary Research* 43: 738–739.

10. *Spatafora, G.A. and T.R. Platt. 1982. Survey of the helminth parasites of the rat, *Rattus norvegicus*, from Maymont Park, Richmond, Virginia. *Virginia Journal of Science* 33: 3–6.

11. Platt, T.R. 1982. Ecological and evolutionary aspects of the helminth fauna of the snapping turtle, *Chelydra serpentina* L., in the United States. *Molecular and Biochemical Parasitology Supplement* pg. 352.

12. Platt, T.R. 1983. Redescription of *Capillaria serpentina* Harwood, 1932, (Nematoda: Trichuroidea) from freshwater turtles in Virginia. *Canadian Journal of Zoology* 61: 2185–2189.

13. Platt, T.R. 1984. Cladistics: A Brief Review. *In* Brooks, D.R., J.N. Caira, T.R. Platt and M.H. Pritchard (eds.), *Principles and Methods of Phylogenetic Systematics: A Cladistics Workbook*, Museum of Natural History: University of Kansas, p. 92.

14. Platt, T.R. 1984. Evolution of the Elaphostrongylinae (Nematoda: Metastrongyloidea) parasites of cervids. *Proceedings of the Helminthological Society of Washington* 51: 196–204.

15. Platt, T.R. and W.M. Samuel. 1984. Mode of entry of first-stage larvae of *Parelaphostrongylus odocoilei* (Nematoda: Metastrongyloidea) into four species of terrestrial gastropods. *Proceedings of the Helminthological Society of Washington* 51: 205–207.

16. Samuel, W.M., T.R. Platt and S.M. Knispel. 1985. Gastropod intermediate hosts and transmission of *Parelaphostrongylus odocoilei*, a muscle-inhabiting nematode of mule deer, *Odocoileus h. hemionus*, in Jasper National Park, Alberta. *Canadian Journal of Zoology* 63: 928–932.

17. *Platt, T.R. and G.A. Harris. 1986. An examination of the prepatent period and the absence of a crowding effect in *Angiostrongylus cantonensis* in laboratory rats. *Journal of Tropical Medicine and Hygiene* 85: 536–541.

18. Platt, T.R. 1988. *Hapalorhynchus brooksi* sp. n. (Trematoda: Spirorchiidae) from the snapping turtle (*Chelydra serpentina*), with notes on *H. gracilis* and *H. stunkardi*. *Proceedings of the Helminthological Society of Washington* 55: 317–323.

19. Platt, T.R. 1988. Phylogenetic analysis of the North American species of the genus *Hapalorhynchus* Stunkard, 1922 (Trematoda: Spirorchiidae), blood-flukes of freshwater turtles. *Journal of Parasitology* 74: 870–874.

20. Platt, T.R. 1988. *Hapalorhynchus beadlei* Goodman, 1987 (Trematoda, Digenea): proposed replacement of the holotype by a lectotype. *Bulletin of Zoological Nomenclature* 45: 258–259.

21. *Ludlam, K.E. and T.R. Platt. 1989. The relationship of park maintenance and accessibility to dogs to the presence of *Toxocara* spp. ova in the soil. *American Journal of Public Health* 79: 633–634.

22. Platt, T.R. 1989. Gastropod intermediate hosts of *Parelaphostrongylus tenuis* (Nematoda: Metastrongyloidea) from Northwestern Indiana. *Journal of Parasitology* 75: 519–523.

23. Platt, T.R. and A.K. Prestwood. 1990. Deposition of type and voucher material from the helminthological collection of Elon E. Byrd. *Systematic Parasitology* 16: 27–34.

24. Platt, T.R. 1990. *Aphanospirorchis kirki* n. gen., n. sp., (Digenea: Spirorchidae) a parasite of the painted turtle, *Chrysemys picta marginata*, from northwestern Indiana, with comments on the proper spelling of the family name. *Journal of Parasitology* 76: 650–652.

25. Platt, T.R., D. Blair, J. Purdie and L. Melville. 1991. *Griphobilharzia amoena* n. gen., n. sp. (Digenea: Schistosomatidae), a parasite of the freshwater crocodile, *Crocodylus johnstoni* (Reptilia; Crocodylidae) from Australia, with the erection of a new subfamily, Griphobilharziinae. *Journal of Parasitology* 77: 65–68.

26. Platt, T.R. 1991. Notes on the genus *Hapalorhynchus* Stunkard, 1922 (Digenea: Spirorchidae) from African turtles. *Transactions of the American Microscopical Society* 110: 182–184.

27. Platt, T.R. 1992. A phylogenetic and biogeographic analysis of the genera of Spirorchinae (Digenea: Spirorchidae) parasitic in freshwater turtles. *Journal of Parasitology* 78: 616–629.

28. *Platt, T.R., D.M. Sever and V.L. Gonzalez. 1993. First report of the predaceous leech *Helobdella stagnalis* (Rhynchobdellida: Glossiphoniidae) as a parasite of an amphibian, *Ambystoma tigrinum* (Amphibia: Caudata). *American Midland Naturalist* 129: 208–210.

29. Platt, T.R. 1993. Taxonomic revision of *Spirorchis* MacCallum, 1919 (Digenea: Spirorchidae). *Journal of Parasitology* 79: 337–346.

30. Platt, T.R. and S. Pichelin. 1994. *Uterotrema australispinosa* n. gen, n. sp. (Digenea: Spirorchidae), a parasite of a freshwater turtle *Emydura macquarii* from southern Queensland, Australia. *Journal of Parasitology* 80: 1008–1011.

31. *Platt, T.R. and S.D. Villanueva. 1995. Postmortem migration of *Hymenolepis diminuta* (Cestoidea: Cyclophyllidea) in the laboratory rat. *Journal of Parasitology* 81: 1024–1027.

32. Platt, T.R. and D. Blair. 1996. Two new species of *Uterotrema* (Digenea: Spirorchidae) parasitic in *Emydura krefftii* (Testudines: Chelidae) from Australia. *Journal of Parasitology* 82: 307–311.

33. Platt, T.R. and D. Blair. 1996. *Hapalotrema* Looss, 1899 (Digenea): proposed designation of *H. loossi* Price, 1934 as the type species. *Bulletin of Zoological Nomenclature* 53: 1–3.

34. Platt, T.R. and D.R. Brooks. 1997. Evolution of the schistosomes (Digenea: Schistosomatidae): the origin of dioecy and colonization of the venous system. *Journal of Parasitology* 83: 1035–1044.

35. Platt, T.R. and D. Blair. 1998. Redescription of *Hapalotrema mistroides* (Monticelli, 1896) and *Hapalotrema synorchis* Luhman, 1935 (Digenea: Spirorchidae), with comments on other species in the genus. *Journal of Parasitology* 84: 594–600.

36. Jue Sue, L. and T.R. Platt. 1998. Redescription and life-cycle of *Sigmapera cincta* Nicoll, 1918 (Digenea: Plagiorchiidae) a parasite of Australian freshwater turtles. *Systematic Parasitology* 39: 223–235.

37. Jue Sue, L. and T.R. Platt. 1998. Description and life-cycle of two new species of *Choanocotyle* n.g. (Trematoda: Digenea), parasites of Australian freshwater turtles, and the erection of the family Choanocotylidae. *Systematic Parasitology* 41: 47–61.

38. Jue Sue, L. and T.R. Platt. 1999. Description and life-cycle of *Thrinascotrema brisbanica* n.g., n.sp. (Digenea: Plagiorchiida), a parasite of the freshwater turtle *Emydura latisternum* from Australia, and the erection of the family Thrinascotrematidae. *Systematic Parasitology* 43: 217–227.

39. Jue Sue, L. and T.R. Platt. 1999. Description and life-cycle of three new species of *Dingularis* n.g. (Digenea: Plagiorchiida) parasites of Australian freshwater turtles. *Systematic Parasitology* 43: 175–207.

40. Platt, T.R. 2000. Helminth parasites of the western painted turtle, *Chrysemys picta belli* (Gray), including *Neopolystoma elizabethae* n. sp. (Monogenea: Polystomatidae) a parasite of the conjunctival sac. *Journal of Parasitology* 86: 815–818.

41. Platt, T.R. 2000. *Neopolystoma fentoni* n. sp. (Monogenea: Polystomatidae) a parasite of the conjunctival sac of freshwater turtles in Costa Rica. *Memórias do Instituto Oswaldo Cruz* 95: 833–837.

42. Platt, T.R. and D.R. Brooks. 2001. Description of *Buckarootrema goodmani* n.g., n.sp. (Digenea: Pronocephalidae), a parasite of the freshwater turtle *Emydura macquarii* (Gray, 1831) (Pleurodira: Chelidae) from Queensland, Australia and a phylogenetic analysis of the genera of the Pronocephalidae Looss, 1902. *Journal of Parasitology* 87: 1115–1119.

43. Platt, T.R. 2002. Chapter 53: Spirorchiidae. *In* Gibson, D.I., R.A. Bray, and A. Jones (eds.), *A key to the digenetic trematodes of vertebrates*, CAB Press: New York, pp. 453–467.

44. Platt, T.R. and R.J. Jensen. 2002. *Aptorchis aequalis* Nicoll, 1914 (Digenea: Plagiorchiidae) is a senior synonym of *Dingularis anfracticirrus* Jue Sue and Platt, 1999 (Digenea: Plagiorchiidae). *Systematic Parasitology* 52: 183–191.

45. Platt, T.R. 2003. Description of *Auriculotrema lechneri* n. gen., n. sp. (Digenea: Choanocotylidae), a parasite of freshwater turtles (Testudines: Pleurodira: Chelidae) from Queensland, Australia. *Journal of Parasitology* 89: 141–144.

46. Platt, T.R. and V.V. Tkach. 2003. Two new species of *Choanocotyle* Jue Sue and Platt, 1998 (Digenea: Choanocotylidae) from an Australian freshwater turtle (Testudines: Pleurodira: Chelidae). *Journal of Parasitology* 89: 145–150.

47. Platt, T.R. 2006. First report of *Echinochasmus* sp. From the snapping turtle (*Chelydra serpentina* L.) from Reelfoot Lake, Tennessee, U.S.A. *Comparative Parasitology* 73: 161–164.

48. Gibbons, L.M. and T.R. Platt. 2006. Three species of nematodes for the family Atractidae (Cosmocercoidea) from *Rhinoclemmys pulcherrima* in Costa Rica. *Journal of Helminthology* 80: 333–340.

49. Platt, T.R. and S.D. Snyder. 2007. Redescription of *Hapalorhynchus reelfooti* Byrd, 1939 (Digenea: Spirorchiidae) from *Sternotherus odoratus* (Latreille, 1801). *Comparative Parasitology* 74: 31–34.

50. *Stillson, L.L. and T.R. Platt. 2007. The crowding effect and morphometric variability in *Echinostoma caproni* (Digenea: Echinostomatidae) from ICR mice. *Journal of Parasitology* 93: 242–246.

51. Zelmer, D.A. and T.R. Platt. 2008. Structure and similarity of helminth communities of six species of Australian turtles. *Journal of Parasitology* 94: 781–787.

52. Rigby, M.C., R.S.K. Sharma, R.F. Hechinger, T.R. Platt and J.C. Weaver. 2008. Two new species of *Camallanus* (Nematoda: Camallanidae) from freshwater turtles in Queensland, Australia. *Journal of Parasitology* 94: 1364–1370.

53. Platt, T.R. 2009. The course of a 300 metacercarial infection of *Echinostoma caproni* (Digenea: Echinostomatidae) in ICR mice. *Comparative Parasitology* 76: 1–5.

54. *Platt, T.R., L. Burnside and L. Bush. 2009. The role of light and gravity in the experimental transmission of *Echinostoma caproni* (Digenea:

Echinostomatidae) cercariae to the second intermediate, *Biomphalaria glabrata* (Gastropoda: Pulmonata). *Journal of Parasitology* 95: 512–516.

55. Zelmer, D.A. and T.R. Platt. 2009. Helminth infracommunities of the Common Snapping Turtle (*Chelydra serpentina serpentina*) from Westhampton Lake, Virginia. *Journal of Parasitology* 95: 1552–1554.

56. *Platt, T.R., H. Greenlee and D. Zelmer. 2010. The interaction of light and gravity on the transmission of *Echinostoma caproni* (Digenea: Echinostomatidae) cercariae to the second intermediate host, *Biomphalaria glabrata*. *Journal of Parasitology* 95: 325–328.

57. *Platt, T.R., E. Graf, A. Kammrath and D. Zelmer. 2010. Diurnal migration of *Echinostoma caproni* (Digenea: Echinostomatidae) in ICR mice. *Journal of Parasitology* 96: 1072–1075.

58. *Platt, T.R., A. Firth and R.S.K. Sharma. 2011. Redescription of *Neopolystoma liewi* Du Preez and Lim, 2000 (Monogenea: Polystomatidae) from *Cuora amboinensis* (Testudines: Geoemydidae) with Notes on Specimen Preparation. *Comparative Parasitology* 78: 286–290.

59. Verneau, O., C. Palacios, T.R. Platt, M. Alday, E. Billard. J.-F. Alliene, C. Basso and L.H. Du Preez. 2011. Invasive species threat: parasite phylogenetics reveals patterns and processes of host-switching between non-native and native captive freshwater turtles. *Parasitology* 138: 1778–1792.

60. *Platt, T.R. and R.M. Dowd. 2012. Age-related change in phototaxis by cercariae of *Echinostoma caproni* (Digenea: Echinostomatidae). *Comparative Parasitology* 79: 1–4.

61. Abebe, E., J. Sharma, M. Mundo-Ocampo and T.R. Platt. 2012. *Testudinema gilchristi n.* gen., n. sp. (Nematoda: Monhysteridae) from the perianal folds of stinkpot turtles, *Sternotherus odoratus* (Testudines), USA. *Nematology* 14: 709–721.

62. Platt, T.R. and R.S.K. Sharma. 2012. Two New Species of *Hapalorhnychus* (Digenea: Spirorchiidae) from Freshwater Turtles (Testudines: Geoemydidae) in Malaysia. *Comparative Parasitology* 79: 202–207.

63. Tkach, V.V., T.R. Platt and S.E. Greiman. 2012. *Opisthioglyphe sharmai* n. sp. (Digenea: Telorchiidae) from gall bladder of turtles in Malaysia. *Journal of Parasitology* 98: 863–868.

64. *Platt, T.R., G.L. Hussey and D.A. Zelmer. 2013. Circadian egg production by *Echinostoma caproni* (Digenea: Echinostomatidae) in ICR Mice. *Journal of Parasitology* 99: 179–182.

65. *Platt, T.R., G. Quintana, A.E. Rodriguez and D.A. Zelmer. Migratory response of *Echinostoma caproni* (Digenea: Echinostomatidae) to feeding by ICR Mice. *Journal of Parasitology* 99: 247–249.

66. Platt, T.R, E.P. Hoberg and L.A. Chisholm. 2013. On the morphology and taxonomy of *Griphobilharzia amoena* Platt and Blair, 1991 (Schistosomatoidea), a dioecious digenetic tematode parasite of the freshwater crocodile, *Crocodylus johnstoni*, in Australia. *Journal of Parasitology* 99: 888–891.

67. Platt, T.R and D.A. Zelmer. 2016. Effect of infection duration on habitat selection and morphology of adult *Echinostoma caproni* (Digenea: Echinostomatidae) in ICR mice. *Journal of Parasitology* 102: 37–41.

68. Roberts, J.R., T.R. Platt, R. Orélis-Ribeiro and S.A. Bullard. 2016. New genus of blood fluke (Digenea: Schistosomatoidea) from Malaysian freshwater turtles (Geoemydidae) and its phylogenetic position within Schistosomatoidea. *Journal of Parasitology* 102: 451–462.

69. *Platt, T.R, F. Gifford and D.A. Zelmer. 2016. The role of light and dark on the dispersal and transmission of *Echinostoma caproni* (Digenea: Echinostomatidae) cercariae. *Comparative Parasitology* 83: 197–201.

70. Platt, T.R. 2017. The Genus *Spirorchis* MacCallum, 1918 (Digenea: Schistosomatoidea) and the early history of Parasitology in the United States. *Journal of Parasitology* 102: 407–420.

71. Bullard, S.A., J.R. Roberts, M.B. Warren, H.R. Dutton, N.V. Whelan, C.F. Ruiz, T.R. Platt, V.V. Tkach, S.V. Brant and K.M. Halanych. 2019. Neotropical turtle blood flukes: Two new genera and species from the Amazon Basin with a key to genera and comments on marine-derived lineages in South America. *Journal of Parasitology* 105: 497–523.

72. Lacmichhane-Khadka, R., A. Slusser, M. Green, D.A. Zelmer and T.R. Platt. 2021. Effect of *Echinostoma caproni* on presumptive lactic acid bacteria abundance and *Salmonella enterica* serovar Typhimurium colonization in the mouse gut. *Journal of Parasitology* 107: 381–387.

Acknowledgments

I want to thank Dr. Phua Kok Khoo and Ms. Yubing Zhai, World Scientific Publishing, for saying "Yes" to *Small Science*. Shaun Tan Yi Jie ably and patiently shepherded this book through the publishing process and deserves my sincerest gratitude.

Ken and Jane Platt loved each other and were married for 55 years until my mother's death in 1996. They believed in their children, and that education was key to a successful life, even if the aptitude didn't manifest itself early on. Their love and support made the events I shared in these pages possible.

Drs. Christopher Hamlin (University of Notre Dame) and John J. Janovy (University of Nebraska) commented on earlier versions of this work. I thank them for their advice and encouragement. I, alone, am responsible for the accuracy of the contents.

The book you have just completed is an extended acknowledgment of the people who influenced my life for the better (again, except for Colleen and D. A. Azimov), and their names need not be repeated here. You know who you are, and I thank you.

About the Author

Dr. Thomas R. Platt is Professor Emeritus at Saint Mary's College, Notre Dame, Indiana, USA, where he was Chair of the Department of Biology from 2001–2008. He is the author of more than 70 publications on parasitology and taxonomy, and the discoverer of 30 new species. He named one of the last parasites he discovered before retiring *Baracktrema obamai*, after then-USA President Barack Obama, who is his fifth cousin, twice-removed. He holds a PhD in Zoology from the University of Alberta, Canada. He may be contacted at trpsmallscience@gmail.com.